Technology and the Multinationals

Technology and the Multinationals

Corporate Strategies in a Changing World Economy

Jack Baranson
Developing World Industry and
Technology, Inc.

Lexington Books
D.C. Heath and Company
Lexington, Massachusetts
Toronto

Library of Congress Cataloging in Publication Data

Baranson, Jack.
    Technology and the multinationals.

    Includes index.
    1. Technology transfer. 2. International business enterprises. I. Title.
T174.3.B364   338.4'5   77-14699
ISBN 0-669-02021-4

Third printing, May 1980

Published simultaneously in Canada.

Printed in the United States of America.

International Standard Book Number: 0-669-02021-4

Library of Congress Catalog Card Number: 77-14699

Contents

List of Tables

Preface

U.S. corporations traditionally have avoided the release of industrial technology, particularly unique and proprietary technology, based on the assumption that the sharing of such technology would weaken and ultimately erode the firm's competitiveness in world markets. Concomitant with this view was the conviction that a return on investment was earned only when a product was developed and successfully commercialized in the marketplace. Retention of the technology on which the product was based assured continued returns throughout the product's life cycle. It was only late in a product's life cycle—once the production techniques had become generally available—that the corporation was prepared to release the technology. Then its release was through the traditional transfer modes of direct foreign investment and licensing of patents and trademarks.

In recent years, there has been mounting evidence that U.S. corporations have begun to depart from this traditional approach toward managing technological assets. Under certain circumstances, a growing number of U.S. corporations now find attractive the sale of industrial technology to noncontrolled foreign enterprises—a phenomenon we shall call "technology sharing." The technology sold in such cases is increasingly the most sophisticated and latest generation available, and its release is often under terms that assure rapid and efficient implantation of an internationally competitive productive capability.

The new approach to world marketing and production implicit in these new technology-sharing arrangements between U.S. corporations and noncontrolled foreign enterprises did not occur overnight nor in response to an isolated event. It is the result, rather, of evolutionary trends in the world economy that have changed corporate viewpoints about foreign involvements and have altered the relative bargaining position of technology purchasers. The combined result of these trends has been to move segments of the U.S. corporate world into what may be termed "commonwealth" arrangements with "sovereign" enterprise affiliates—and away from traditional "empire" relationships with owned and controlled subsidiaries. The major factors that have influenced trends in this direction are:

1. The demands of newly industrializing nations for technology sharing and access to world markets
2. The intensified political risks and economic uncertainties of overseas capital investments in plant and equipment
3. The shifting emphasis in certain firms from production to marketing and R and D functions
4. The intensified competition from foreign enterprises as suppliers of

industrial technology and the consequent compulsion to release proprietary technology early in the product cycle
5. The escalation of R and D and capital investment costs connected with
the proliferation of world involvements and the ever-increasing
sophistication of product systems

This book elaborates on the foregoing themes and draws upon a series
of case studies of international transfers of industrial technology by U.S.
firms that were carried out between June 1975 and October 1976 under a
research grant from the Office of Foreign Economic Research, Bureau of
International Labor Affairs, U.S. Department of Labor. Our original
research objective was to develop policy implications for the U.S. economy,
but the case material also contains important implications for corporate
management of global operations and for the development strategies of
newly industrializing nations.

The introductory chapter describes the new international setting that
has lead to the type of technology-sharing arrangements described in the
case material. In addition to revealing emerging trends in the world
economy that have compelled major changes in corporate global involvements, it also identifies and illustrates the changing strategies adopted
by U.S. technology suppliers and foreign-enterprise purchasers. Chapter 1
closes with a note on the case method, the analytical framework, and definitions.

The case studies, contained in Chapters 2-6, are classified by industrial
sector—aircraft, automotive, computer, consumer electronics, and
chemical engineering—and focus on the release of unique and proprietary
technology to noncontrolled foreign enterprise. Each chapter of case studies
starts with a sector overview, which examines the characteristics and recent
trends in the industry that bear on corporate decisions to release technology
to noncontrolled foreign enterprise. Chapter 7 sets forth the policy implications for corporate management of global operations, foreseeable impact
upon the U.S. economy, and implications for industrialization in developing world nations.

I am most grateful for the substantial contributions made to this book
by my associates and colleagues in the worlds of business, academia, and
government. I should especially like to acknowledge the support and
guidance leant this effort by Harry Grubert of the U.S. Department of
Labor. He and others who sat through the review sessions offered critical
and insightful comment throughout the field research and the drafting of
case studies and the final analytical sections.

The case writers involved in this project provided excellent material on
which to base my analysis. Their extensive knowledge of U.S. industry as
well as their familiarity with the technical aspects of the work greatly eased

the burden of sifting through and assimilating highly technical data. These individuals were: Richard M. Bissell of The Exchange; Y. S. Chang, Boston University; Robert H. Fuhrman, consultant; Peter A. Glenshaw of Glenshaw, Inc.; Anne Harrington of DEWIT; and John H. Hoagland of Hoagland, MacLachlan, Inc.

I am especially grateful to Anne Harrington, who shared the burden of writing and editing from the inception of the project to its final stages of production.

Numerous corporate officials gave generously of their time and ideas and in several instances served as my most constructive critics. Industry specialists with trade associations were also an extremely useful source of information.

I am most grateful for the funding provided by the Office of Foreign Economic Research, International Labor Affairs Bureau, U.S. Department of Labor, as well as for its continued interest in and support of my work. The views and opinions expressed in this book, however, do not necessarily represent the official position or policy of the Department of Labor. I am also grateful to the editors of *Foreign Policy* who have permitted me to use material that appeared in an article I wrote for the Winter 1976-1977 issue of that journal. This material appears largely in Chapter 1.

I am hopeful that the study will contribute to a more perceptive and effective dialogue between government policy makers and managers of U.S. industry in matters concerning the international transfer of U.S. technology. The case material indicates some profound changes and shifts in the foreign demand for technology and in the technology packages that U.S. firms are now supplying to foreign enterprises and suggests that these changes may have profound implications for necessary adjustments in the U.S. economy. The case materials also point to a broad range of opportunities, heretofore denied, that are opening up to the developing world nations to acquire industrial technology on more advantageous terms.

Technology and the Multinationals

1 The New International Setting for Technology Sharing

Certain trends in the world economy have compelled U.S. corporations to alter their mode of foreign involvement and to revise policies governing management of their technological assets. In these changing circumstances, U.S. firms are finding new opportunities to earn returns on their managerial and technological assets. In some instances, the "product" has become the implanting of design and engineering capabilities that are the spawning grounds of future industrial competitors. Technology purchasers likewise face new sets of options and opportunities for acquiring foreign know-how and usually on more favorable terms. This chapter explores the relevant changes that have occurred in the global environment, the considerations that have influenced corporate response to these changes, and the motivations that have led to new strategies by technology purchasers.

Changes in Corporate Perspectives

From the corporate perspective, the attractions of equity investment in and managerial control of foreign facilities are waning. A growing number of U.S. firms have decided that the risks associated with overseas capital investments have become too high for realized rates of return. Aside from political uncertainties in a widening area of the world, there are economic vicissitudes brought on by world inflation, exchange rate revaluations, and recessionary cycles, all of which have added to the risks of locking into fixed investments in a world of changing circumstances. These uncertainties have been compounded by the fragmentation of world markets resulting from import substitution behind tariff barriers and regional trading blocs as a partial offset to the inefficiencies of protected national economies.

Although most U.S. corporations earned profits overseas in 1976, overall return on investment did not meet expectations. According to the U.S. Department of Commerce, profits from foreign operations of U.S. corporations rose 13 percent in 1976 to $18.86 billion from $16.62 billion in 1975. This figure fell short, however, of the 1974 high of $19.16 billion. The changes that took place in currency values in 1977 largely tended to reduce profits of U.S. concerns, especially those operating in Mexico, Brazil, and other countries of Latin America. In response to this declining profitability, U.S. firms are investing abroad with greater caution. Net capital outflows, for example, declined 27 percent in 1976 to $4.6 billion.[1]

1

The growth of host country restrictions, regulations, and limitations on foreign investment has also tended to detract from this traditional mode of overseas involvement. Import restrictions are commonly imposed by countries that, like Brazil, suffer balance-of-payments deficits due to increasing oil import bills. Even some of the oil-rich countries, such as Venezuela, keep a close eye on imports and costly foreign-exchange expenditures, however. Another inhibiting factor in this area is the limitation by law of employee layoffs. These laws in some countries are so strict that a U.S. corporation is forced to consider labor a fixed cost, regardless of prices or demand for the product. In addition to being legally obligated to make individual termination payments, some U.S. corporations must also develop, in collective bargaining with unions (and sometimes with the government), "social plans," which require additional cash indemnities to workers.

At the same time that certain changing circumstances in host environments have soured corporate attitudes toward direct investments in overseas plants, there is some evidence that U.S. firms are encountering real difficulties in adjusting to technical change at home and aggressively engineering for competitive production in the high-wage U.S. economy. Offshore production of the labor-intensive component of manufacturing for reexport back to the U.S. for final assembly has long been a practice of U.S. industry. Today, however, some of the more capital-intensive elements (such as entire engine plant facilities) are allocated to foreign affiliates in order to maintain competitiveness in world markets.

U.S. firms are also retrenching from new investment commitments in basic research aimed at new product development in favor of pragmatic, quick payoff development of existing technology. The reorientation in research and development (R and D) coincides with a virtual stagnation in R and D spending by U.S. corporations. The increase in expenditures of this kind in 1976 was barely commensurate with the rate of inflation that year. Explanations for the reorientation in R and D by U.S. corporations include continued high rates of inflation, shortage of capital funds due to the stock market slump, fierce competition worldwide in high-technology industries, and uncertainty in U.S. government regulations and policies. An even more ominous development is that the share of R and D done by the U.S. has begun to slacken compared with the rest of the world. One government study estimates that only about 20 percent of the world's R and D is currently done by the U.S.—down from 50 percent a mere decade ago.[2]

An increasingly compelling alternative to either overseas or domestic investment in production facilities and R and D has emerged in the sale of manufacturing know-how and related services, such as training of foreign nationals in technical and managerial aspects of operating industrial systems or assistance in procurement, startup, and, most important, international marketing.

Enhanced Bargaining Leverage of
Technology Purchasers

The degree of autonomy U.S. corporations have traditionally enjoyed in making decisions as to the kinds of technology they released and the terms of transfer has been considerably diminished by a gradual increase in the bargaining leverage of technology purchasers. The enhanced bargaining leverage derives from a variety of new conditions, which redound to the advantage of foreign purchasers. Any one condition, in and of itself, is not necessarily determining; but combined these new conditions result in a marked shift in the balance of bargaining leverage between technology suppliers and purchasers in favor of the latter.

The shift in bargaining leverage is perhaps most immediately attributable to the proliferation of alternative sources of similar, if not equally attractive, technology sought by a purchaser. The number of product and process fields in which the U.S. industry holds a monopoly position in know-how is rapidly declining. That enterprises in other industrially advanced countries have become alternative suppliers of first-class, operative technology is a result, in part, of an earlier proliferation of design and manufacturing capabilities by U.S. firms. The Japanese, for example, have become significant competitors with U.S. technology suppliers, because they have absorbed and adapted segments of U.S. technology in the late 1950s and 1960s. The socialist countries may follow suit in time, depending on the volume of transfers and the efficiency of implantation.

Not only does foreign competition exist in a wide range of industrial technological fields, but the foreign suppliers in many cases are characterized by astuteness and aggressiveness in selling the know-how. A reluctance to tailor the terms of transfer to accommodate the demands or needs of the purchaser is not nearly so evident among foreign technology suppliers. The tortuous phase of soul-searching with which many U.S. firms are now grappling concerning how best to manage corporate technological assets and to respond to the new set of demands being placed on them by foreigner technology purchasers was passed through relatively painlessly by their foreign competition.

The supportive and interventionist role played by the governments of foreign technology purchasers has also greatly expanded in recent years, thereby lending increased bargaining leverage and authority to the purchasing enterprise in negotiating with American corporations. The precise nature of state support or intervention varies from one country to another and from one commercial transaction to another, but national economic development authorities now recognize the link between government involvement and industrial growth that is compatible with and contributes to articulated national objectives.

As the case materials illustrate, the expanded role of government assumes several forms, ranging from subsidization of R and D in joint ventures between national and foreign firms (typical of industrially advanced country governments) to acting as prime negotiator and purchasing agent in technology transactions (a widespread practice by socialist and, to a lesser extent, by newly industrializing countries' governments). Most foreign governments have resorted to protectionism of one variety or another as a means to compel U.S. corporations to transfer technology to industrial facilities established within protected markets or to give up those markets completely. A variety of other forms of support is provided foreign enterprises by their governments, ranging from the subsidizing of bidding costs for overseas contracts and cost overruns of overseas jobs to the collection of commercial intelligence in foreign countries and its dissemination to the private sector.

Another strong bargaining factor is that many of the socialist governments are prepared to provide investment capital and pay for design and engineering costs—an attractive inducement to U.S. firms suffering capital shortages to share manufacturing know-how. To those U.S. firms that are loathe to alter traditional investment and licensing patterns, the expanded support and intervention of foreign governments in the negotiation process are deeply resented; as the case studies indicate, however, an increasing number of U.S. corporations are anxious to exploit to their own advantage selected forms of foreign government assistance to national enterprise.

Developing nations with oil and mineral resources have begun to use their newly derived wealth to acquire industrial facilities and know-how to process their raw materials for world markets. Most of these resource-rich countries have heretofore played marginal roles in the world economy. Their greatly enhanced earnings obtained from coordinated pricing and volume-of-production policies have given them considerable bargaining leverage in negotiating with multinational corporations. This kind of leverage has accrued not only for the oil-rich countries but also to a lesser extent for nations having bauxite, copper, and other raw materials. The ability to pay in hard currency for expensive technology, which eliminates complicated and long-term financing arrangements, is being used as strong leverage in negotiations.

A final factor in the shift in bargaining leverage has been the perceptible upgrading in recent years of the knowledge and skills foreign technology purchasers bring to bear in negotiation sessions with multinational corporations. Trained in the best business and law schools in the world, negotiators for purchasing enterprises have become well versed with the laws concerning industrial property rights, patents, trademarks, and industrial know-how.

U.S. Corporate Response to Changes in
the World Economy

The case material reveals at least four strategies that explain the willingness of U.S. corporations to release proprietary technology to noncontrolled foreign enterprises. Perhaps the most dramatic of the new strategies has been the adoption by some U.S. firms of an explicit policy to shift from equity investment and managerial control of overseas facilities to the sale of technology and management services as a direct means to earn returns on corporate assets. The decision to release corporate assets that have traditionally been considered of the highest proprietary value is part of a larger and more comprehensive new logistic of international marketing and production devised by several multinational firms. Consider the following arrangement.

Cummins Engine Company, a billion-dollar corporation, is committed to sharing the production function on its newest generation of diesel engine design with its longstanding licensee in Japan, Komatsu, a leading tractor and construction equipment manufacturer. Based upon a comparative analysis of capital requirements for future expansion of R and D, marketing, and production capacities, the latter was found to be least efficient in terms of relative rates of return. Cummins' long-range planning seems to point toward a shifting center of gravity in the direction of a technology and marketing company, rather than a manufacturing company.

The second strategy that emerged from the case studies is closely related to and, in fact, is intertwined with the preceding one. In this case, the U.S. corporation is prepared to release proprietary technology to a foreign enterprise (either through a joint venture or divestiture) in order to alleviate capital shortages, to offset the enormity of R and D or production-tooling expenditures, or to circumvent nontariff barriers. The commonality of the first and second strategies consists of the corporate philosophy or attitude that permits the sharing of highly sophisticated, proprietary know-how; the distinction lies in the exigency that compels the corporation to enter into the arrangement; that is, in the absence of adverse conditions, the technology would in all probability be retained under corporate auspices. Needless to say, this distinction frequently becomes blurred, and attribution of a particular corporate action to one of these two strategies must be arbitrary in some cases.

U.S. firms are particularly vulnerable where they face the necessity to accept a foreign affiliate due to prohibitively high R and D costs or production-tooling requirements. Nontariff barriers to entry or offset purchase requirements—typical of the aircraft industry—are two other conditions that undermine the U.S. firm's bargaining position, particularly where the foreign enterprise is strongly supported by government policies and financial resources. Honeywell's approach to the French market exemplifies this particular strategy.

In 1975, Honeywell sold a portion of its ownership in its French subsidiary, Honeywell-Bull (HB), in order to assume a minority equity position and, thereby, to become eligible for a merger with Compagnie Internationale pour l'Informatique (CII), the French government-sponsored computer firm. Whereas Honeywell had previously been excluded from the French public sector market, it was then able to increase substantially its sales and earnings in the French market and was able to enjoy access to the very significant R and D subsidies offered by the French government. The French obtained majority ownership of HB, management control of the entity known as HB-CII, and, most important, access to an ongoing flow of advanced U.S. computer technology.

A third corporate strategy encountered in the case studies entails what may be called the measured release of core technology. This strategy is most common within the process design and engineering industry where the company is interested in selling newly designed technology as extensively as it can and reinvests a portion of profits in developing new generations of technology. In most cases, a deliberate attempt is made to retain an essential element of the know-how, without which the purchasing enterprise is unable to develop a more competitive version or to become self-sufficient in the technology.

In cases involving the measured release of core technology, much depends on the astuteness and bargaining position of the foreign purchaser relative to the U.S. firm. The ability of the U.S. firm to maintain its technological lead and to continue to commercialize that lead effectively in world markets also contributes to the success with which a corporation employs this strategy, its ultimate effects on the U.S. economy, and its implications for long-range corporate planning. The Fluor Corporation, one of the world's largest chemical engineering companies, has thus far employed this strategy in its overseas design-engineering contracts with considerable success.

Fluor's major operations are in the engineering and construction of chemical and petroleum facilities. The company has achieved some advantages over its competitors by converting many of the management aspects of new plant design and construction into mathematically programmable routines—the substance of Fluor's project management technology. It has developed its technology of project management into an art, which, to a greater or lesser degree, is transferred every time it performs on an engineering and construction contract. It delivers some of the techniques only as part of the total responsibility package that it supplies. It would be difficult, if not impossible, for a purchaser of these discrete parts to reproduce the technique in its entirety.

A final strategy that emerged from the case studies entails a corporate decision to earn returns on technological assets that are not considered central to the company's business or are no longer commercially viable. Such cases may include the licensing of a standard range of the company's technology or, in extreme cases, divestiture of an entire division. Motorola's arrangement with Matsushita exemplifies this approach.

In 1974, Motorola sold the vast majority of its color TV manufacturing assets and patents to Matsushita Electric Industrial Company, the world's largest consumer electronics manufacturer. Motorola's technology was rapidly becoming obsolete and that line of its business was no longer commercially viable without substantial investment in the R and D and marketing functions. The firm had only 7 percent of the U.S. market for color TVs, and the market was approaching saturation at the time. As a result of the sale, Matsushita, although it did not acquire especially competitive technology, was able to make major inroads into the U.S. market.

The sale of technology that is no longer central to a company's business or commercially viable can be an attractive strategy to the particular firm, but it may also have important backlash effects on other U.S. firms in domestic or foreign markets. A technically strong and commercially astute foreign purchaser may preempt third-country markets, or second-generation R and D by the foreign enterprise may result in a new aggressive outward thrust.

New Strategies of Technology Purchasers

From an analytical viewpoint, it is useful to distinguish among three distinct purchaser environments: industrially advanced economies (Japan and Western Europe); socialist countries (the Soviet Union, the People's Republic of China, and Eastern Europe), and developing world nations that are resource rich (Algeria and Venezuela) and/or have a relatively advanced industrial base (Brazil and Iran). There are substantial differences among and within these groupings in terms of government policies, bargaining power (i.e., financial resources, astuteness in orchestrating contract negotiations, and access to alternative technology sources), and technical absorptive capabilities (at the enterprise level and among supporting industrial sectors).

A basic purpose in purchasing operational technology is to gain time and ultimately to save resources. Newly acquired technology may be used to modernize or otherwise upgrade existing plant and industrial facilities or to move into new or improved product lines and to acquire new or improved production or materials-processing techniques. From the governmental viewpoint, industrial development goals may range from the narrowly economic (to improve productivity or to enhance competitiveness in world markets) to broader developmental and political goals (to develop decreased dependence or absolute self-reliance in basic or defense-oriented industries). In both instances there are important distinctions between turnkey (operational) technology and in-depth training and technical information that would permit a technology purchaser to duplicate acquired technology and to innovate beyond that point.

It is an official policy of several Western European governments and the Japanese government that national enterprises should acquire advanced technologies, particularly in the fields of computers, jet aircraft, and certain automotive and electronic components in order to develop internationally competitive industries. An increasingly popular mode of transfer in these cases is the formation of a joint venture between the U.S. firm and the foreign firm or consortium of firms in which the former provides the technology and the latter, with government financing, provides venture capital and R and D and production-tooling funds or the means by which to circumvent nontariff barriers to market entry. To the U.S. firm suffering severe capital constraints and finding itself excluded from important markets, these can be extremely attractive terms on which to negotiate a joint venture.

Technology-purchaser enterprises in the industrially advanced countries have a high absorptive capacity of latest-generation U.S. know-how and the ability not only to duplicate and innovate U.S. industrial technology but also to more successfully commercialize and market the improved technology. The Amdahl-Fujitsu relationship is a significant case in point.

Between 1972 and 1975, Fujitsu, Ltd., a leading computer manufacturer in Japan, acquired progressive access to highly advanced computer technology. The Amdahl Corporation, which was founded in 1970 by a former IBM design engineer, had developed the technology, which enables users of IBM equipment to double the cost effectiveness of their data processing. In return for successive rounds of venture capital, Fujitsu, backed by Japanese government financing, acquired full patent and manufacturing rights for Japan and has since moved the technology to Spain and is negotiating with Siemens in Germany to share Amdahl technology. Eighty percent of Amdahl computer manufacturing requirements have been shifted to Fujitsu in Japan. Fujitsu is now planning to use Amdahl component technology to design and manufacture small to medium computers for sale in the U.S. market.

The changing role of the technology factor in world trade dynamics is nowhere more evident than in recent developments in the nonmarket economies of Eastern Europe, the Soviet Union, and the People's Republic of China. With the advent of détente and the subsequent expansion of foreign trade and commercial opportunities, many of these countries have seized upon technology as the means to enhance their competitive advantage in world trade. Technology-sharing arrangements with socialist countries have been state negotiated and are aimed at technological self-sufficiency. Poland and Rumania, in particular, have spearheaded the drive to negotiate the new generation of technology-transfer agreements, which involve the implanting of internationally competitive technology and production systems (high-volume, cost-efficient, and quality-controlled products).

The new generation of technology-transfer agreements with the socialist economies is characterized by a unique and, in some cases, unprecedented set of conditions and arrangements. It progresses several stages beyond the so-called "turnkey" agreement, whereby a foreign firm assumes all responsibility for the project from the initial feasibility and cost study phase, through design, engineering, and plant construction, to the final phase of inserting the "black box" technology component into the industrial system. Typically, these new arrangements include: fast and efficient enterprise-to-enterprise implantation of production capabilities; patent and trademark rights; marketing of end products outside socialist country markets; extensive training of nationals; technological upgrading of original know-how sold; and in some cases, the design of a new or improved product line that can compete in world markets, engineered to be produced at competitive cost in the East European environment.

Enterprises in the socialist or nonmarket economies have several well-defined and interrelated objectives in mind when negotiating the new kind of technology-transfer agreements. Perhaps the most compelling objective is to rapidly acquire internationally competitive technology with which to manufacture products previously imported and to increase exports. Import substitution and/or export promotion permit(s) the saving or earning of sorely needed foreign exchange with which to further expand and upgrade industrial production. Earning of hard currency is also critical in light of the tremendous debt burden now carried by these countries. The group's strong interest in obtaining western trademarks is evidence of its desire for acceptance of its products in international markets.

Achieving self-sufficiency in design and engineering capabilities is another important objective of the socialist countries. This ability would enable them to adapt foreign technology to their own factors of production and, more important, to generate their own technologies. Many of the East European countries are eager to reduce their technological dependence not only on the industrially advanced countries but also on the Soviet Union. The following scenario drawn from the case studies illustrates the interrelated set of strategies pursued by this purchasing group.

General Motors Corporation entered into negotiations for a new agreement with a Polish state-owned automotive manufacturing enterprise (POLMOT) that represented a landmark of change in U.S. business involvement in overseas markets. Under the proposed agreement, General Motors would design and engineer for volume production (100,000 units/year) in Poland a new line of vehicles ranging from light pickup vans to medium trucks in the 5-metric-ton range. General Motors would market under its trade name a percentage of the Polish plant's output—15,000 units is the likely number, initially. GM would train Polish managers and technicians in their facilities located in England and other parts of Europe, and Polish technicians would participate in the design and engineering of the new truck line. The proposed GM agreement went a significant step beyond the coproduction and comarketing agreements that have been negotiated by American and other foreign enterprises in Eastern Europe in that it included the sharing of design and engineering of a new product line for the highly sophisticated and competitive Western European market.

The third grouping of purchasers—the advanced or affluent among developing nations—is characterized by articulated government policies of rapid industrialization based on foreign technology and rapid growth of industrial facilities to process and fabricate their raw materials for world markets. The goal of technological self-sufficiency has assumed a central position in many of the industrializing countries' development plans, and the strategy toward that end encompasses the progressive transfer of capabilities to manage and take substantial control over industrial facilities in lead sectors. Increasingly, countries within this group are seeking to reduce the predominant positions of foreign enterprise in their national economies and to wean themselves away from near-exclusive dependence upon multinationals as suppliers of technology. In addition, they want national enterprise to move from the shelter of protected markets, characteristic of the import-substitution phase of industrialization, into the more competitive environments of regional and world export markets—the aim being to improve productivity of resource utilization and balance-of-payments positions.

Implicit in many of the new agreements are motivations on the part of the purchasing authority to lay the base for an internationally competitive industry and an expanding self-reliance to design and engineer future generations of production systems adapted to emerging national needs and industrial capabilities. Two other considerations in contract relations involving the design and construction of industrial facilities in industrializing countries are the training of technicians and engineers to duplicate and innovate and the progressive involvement and development of domestic capital-goods industries and related supplier plants as contributions to overall industrial growth and development. Certain resource-rich developing countries, while low in technological absorptive capabilities, do have the financial resources to purchase comprehensive packages of the best technology available. Consider the following case:

In 1972, General Telephone and Electronics signed a $223 million contract with SONELEC, an Algerian state-owned enterprise, to build a completely integrated consumer electronics plant from raw materials, through component manufacture (including cathode-ray tubes and semiconductors), to end products (TVs, radios, tape recorders, and cassettes). Under the agreement, over 300 Algerian technicians and managers are to be trained in the United States at a cost of $25 million. SONELEC is planning to manage its Sidi-Bel Abbes facility entirely with Algerian nationals.

Table 7-1 gives a summary overview of the case studies by corporate strategy and purchaser objectives.

A Note on the Case Method, Analytical Framework, and Definitions

Several introductory comments concerning methods of collecting information for the case studies and the analytical model employed to derive policy implications follow. The case study approach was selected because it responds to the reality of corporate decision making and it permits focus on a specific technology (many companies are highly diversified and are likely to handle different product lines in different ways). Special emphasis has been directed to areas where U.S. technology is in high demand either because of its technical sophistication or because of the unique package of technical know-how and managerial organization that contribute to efficient and competitive industrial systems.

Selection of the case studies was based on several criteria. First, each case study identifies a particular agreement between a U.S. firm and a non-controlled foreign enterprise that resulted in the transfer of proprietary, industrial technology to the foreign enterprise. Secondly, our analytical model called for a diversity of corporate considerations, purchaser environments, and technology packages. We attempted to select cases that collectively would provide a representative sample of this universe.

Access to and availability of information about a particular agreement were also important determinants in case study selection. The subject of international licensing and the transfer of technology is, in most technology-based companies, highly proprietary. Although general policies and hypothetical scenarios are acceptable topics of discussion, most U.S. corporations are loathe to disclose details of industrial licensing arrangements. There are several reasons for this reluctance. First, there is a pervasive feeling within the U.S. business community that recent public scrutiny of their overseas operations has reduced their flexibility in conducting such operations. Second, several agreements negotiated by U.S. firms in recent years have been highly controversial and have generated heated debate in government and labor circles and in the media on the subject of corporate responsibility beyond the shareholder. Finally, some international licensing agreements can be construed, at times, as having trade-restricting overtones, and U.S. firms, anxious to avoid any suggestion of antitrust violation, are reluctant to discuss particular agreements.

Perhaps, though, the most important explanation of why a firm would decline to participate in a study of this nature is simply the demand on corporate time. Numerous studies have been conducted in recent years on the multinational corporation and they entail long questionnaires and extensive interviews. Combined with the obligatory requests for information from U.S. government agencies, such as the IRS, the Commerce Department, and the Securities and Exchange Commission, these voluntary interviews can become quite onerous.

One of the firms interviewed was granted anonymity in the case report in exchange for providing sensitive material on a particular technology-transfer agreement. In this case, a pseudonym replaces the real name of the firm in the case study. In cases for which relevant information was denied by the U.S. party to the agreement, the case writer had to rely on information obtained from trade journals, officials in trade associations, U.S. government officials with knowledge of the agreement, and academicians. Most of the case studies are based on a combination of interviews with corporate executives and informed noncorporate individuals, and readings in diverse publications ranging from congressional testimony and trade association literature to annual reports, business magazines, and newspapers. In developing scenarios, we were occasionally confronted with the "Rashamon phenomenon"; in other words, sources of information told different stories about the same set of events. When discrepancies appeared, it was left to the case writer's discretion and judgment to resolve or explain the differences in viewpoint.

The case study chapters are divided by industry, and in addition to case studies, each contains an overview of trends within the industry that bear on corporate decisions regarding management of technological assets. A draft copy of each case study was sent to the relevant corporate official (generally the vice president for international operations or international licensing and patents) with a request that it be reviewed for factual or interpretive inaccuracies. We also solicited from the officials additional material or information they felt bore on the case. Significant differences in viewpoint or interpretation are noted.

We developed an analytical model designed to draw implications from the case materials for three policy sets involved in technology-sharing arrangements: U.S. corporations, the purchasing enterprise, and the U.S. economy. The technology transfers depend upon an interrelated set of factors consisting of the motivations, strategies, and capabilities of technology suppliers (conditioned, in part, by government policies and economic situations); the astuteness, bargaining power, and absorptive capabilities of recipient enterprises (as reinforced or conditioned by government action and economic policies); and the nature, quantum, and complexity of the technology transferred. Each of these elements is described in the case scenarios.

A summary view of the analytical model appears in Table 1-1. What is actually transferred and the timing of these transfers are functions of the supplier-enterprise strategies, supplier-government policies, purchaser-enterprise objectives and bargaining position as reinforced or conditioned by purchaser-government policies and activities.

Technology, as used in this study, refers to the package of product designs, production and processing techniques, and managerial systems that are used to manufacture particular industrial products. A modern diesel truck engine consists of approximately 750 individual parts and requires over 30,000 separate steps to convert industrial materials (over 350 different kind) into finished components. The technology package consists of detailed process sheets, materials specifications, processing and testing equipment designations, and quality-control procedures. Together they are a measure of the "quantum and complexity" of a technology transplant.[3] These tens of thousands of elements for a single industrial product, such as a high-speed diesel, are meticulously accumulated over time through research and development, through trial and error in equipment and factory methods, and in the detailed specifications and procedures developed through prolonged experience. The quantum and complexity of technology in the chemical processing and other high technology fields, such as computers and jet aircraft, become even greater.

Table 1-1
Analytical Framework: Technology Transfers by U.S. Corporations

Supplier Enterprise (Government)	Technology	Purchaser Enterprise (Government)

*U.S. government policies:
  Economic consequences
   (erosion of production
   jobs)
  Political-strategic con-
   siderations (transfer of
   industrial capabilities
   in strategic sectors)
 	
* Bargaining power (enter-
   prise):
  Financial resources and
   international experience
  Technological lead and
   world market position
   of enterprise
 	
* Enterprise strategies:[a]
  Shift from equity invest-
   ment and management
   control to sale of
   technology and man-
   agement services
  Necessity to accept for-
   eign affiliate due to
   enormity of R and D
   or capital investment
   costs, offset require-
   ments, or because
   scale of operation re-
   quires consortium
  Measured release of core
   technology
  Sale of technology no
   longer central to com-
   pany business or com-
   mercially advantageous
	*Distinctive characteristics:
	  Quantum and complexity
	  License to manufacture
	   or turnkey-plus
	  Operative-duplicative-
	   innovative
	  General-firm-system
	   specific
	  Stage in product/process
	   cycle
		*Government policies:[a]
		  Industrially advanced nations
		  Socialist countries
		  Developing nations
		
		*Bargaining power:[a]
		  Absorptive capabilities
		  Alternative sources of tech-
		   nology
		  Astuteness
		  Financial resources
		
		* Enterprise strategies:
		  Internationally competitive
		   technology
		  Duplicative and/or innovative
		   design and engineering
		   capabilities
		  Entry into export markets
		  Training of technical mana-
		   gerial manpower
		  Fast, efficient technology
		   transplants

Source: Developing World Industry and Technology, Inc.
[a]See Table 7-1.

Technology transfer is best effected through sustained enterprise-to-enterprise relations, during which time, elements of the technology package are conveyed to the recipient through the provision of the aforementioned technical data, through training and demonstration, and through an interactive process of adjusting implanted procedures and equipment operation until the desired end products, including production inefficiencies and quality standards, are achieved.

The continuum of technology packages may range from mere patent and trademark rights to a complete "turnkey-plus" plant that includes all of the foregoing. Management service contracts stand in the forefront of highly effective technology transfer modes to implant operative technology—including the training of managers and technical supervisors—in a rapid and efficient way.

An important set of distinctions needs to be drawn among "firm-specific," "system-specific," and "general" technology.[4] The first refers to the tried and tested practical knowledge that a firm has developed over time to produce a particular product—with all the bits and pieces fitting together and functioning at a cost-competitive level. "System-specific" refers to specialized capabilities that a firm may have developed over time in such areas as welding techniques (for example, to attach the fins of a turbojet engine to the drive shaft) or casting techniques for a special alloy (of high quality without porosity for the same turbojet engine). "General" knowledge is the easily obtained, nondetailed information about design and manufacturing principles for, say, fractional horsepower electrical motors.

Another critical aspect is the distinction between implanting "operational" (turnkey) technology and imparting technical capabilities to duplicate that technology, which, in some cases, may lead to an indigenous capability to design and engineer industrial systems. The Japanese have been particularly successful in using licensing arrangements as stepping stones to self-sufficient technological capabilities, which were eventually used to develop new generations of internationally competitive products, processes, and production systems. Achieving worthwhile results in the latter area depends, in part, on the stage of development of the recipient enterprise and supporting industrial sectors.

Notes

1. For a discussion of these and other indicators of U.S. foreign investments, see U.S. Department of Commerce, Bureau of Economic Analysis, *Survey of Current Business,* vol. 57 (August 1977), no. 8; and "Tougher Going: U.S. Firms Dissatisfied with Earnings Abroad as Economies Sputter," *Wall Street Journal,* 23 September 1977.

2. Aggregate U.S. expenditures for R and D have slipped to about 2 percent of gross national product, compared to over 3 percent in the mid-1960s. See "Backing Off Basics—Many Concerns Stress Product Development and Reduce Research," *Wall Street Journal,* 18 October 1977.

3. For further details on measuring and quantifying this "quantum and complexity," see Jack Baranson, *Manufacturing Problems in India: The*

Cummins Diesel Experience, Syracuse, N.Y.: Syracuse University Press, 1967, pp. 18-23.

4. This basic distinction is drawn in George A. Hall and Robert E. Johnson, "Transfers of United States Aerospace Technology to Japan," in *The Technology Factor in International Trade,* edited by Raymond Vernon, New York: Columbia University Press, 1970.

2 The Aircraft Industry

Sector Overview

International marketing and production strategies of the U.S. aircraft industry are being significantly modified—in part due to its experiences, but more as a consequence of changing circumstances. In the future, manufacture abroad will be carried out far less frequently under conventional licensing agreements whereby manufacturing rights are granted a foreign firm (or government) in return for royalty payments. What will occur increasingly will be highly organized collaboration in the international manufacture of aircraft among or between companies that agree in advance on their respective share of design, development, and manufacturing costs and activities.

Past Experiences

The more profitable way for U.S. aircraft corporations to exploit foreign markets has been to export U.S.-manufactured products. As a practical matter, however, opportunities to export finished aircraft—at least to industrially advanced country markets—are decreasing rapidly. The government of the purchasing party is generally directly or indirectly involved in such decisions—whether they concern civilian or military aircraft—and it usually has political and military, as well as economic, reasons for insisting on national production. For the most part, therefore, the motives for licensing have been essentially defensive.

Any net benefits derived from licensing have been regarded by the U.S. firm as byproducts of its investment in technology. The benefits sought in granting licenses have been, on occasion, to facilitate the sale of U.S.-produced equipment and, more often, to promote the sale of components and of spares, especially in markets that, for reasons described previously, could not be served by the export of equipment wholly manufactured in the U.S.

Licensing has served as an effective mechanism of transferring technology by the U.S. aircraft firms to noncontrolled foreign enterprises. Although the complexity of the technology is high, a process of more or less natural selection has insured that the licensees to which transfers were made were technically sophisticated enterprises that were capable (at times with some assistance) of absorbing it. Although there are benefits to be gained from conventional licensing agreements to the U.S. firm, such as revenue from royalties and the means to circumvent nontariff barriers to market entry, the potential cost associated with some licensing agreements has become prohibitively high. In the absence of contractual injunction, a licensee or, more commonly, a former licensee independently may further develop the technology licensed and compete with the U.S. licensor or other U.S. aircraft firms in export markets. In one of the two case studies we have done in the industry involving licensing, that of Piper's ten-year agreement with EMBRAER in Brazil, such a scenario is quite likely to materialize.

There are, additionally, reasons why licensing has outlived its usefulness and, therefore, its attraction for potential licensees. One of the major consequences of past licenses has been to allow the aerospace industry of other industrially advanced countries to catch up in their technology in the shortest possible time and at minimum cost. To a large extent, this goal has been reached, and therefore, an important reason for seeking licenses has become less urgent.

The preceding comments on traditional licensing are limited to overseas production in the industrially advanced nations of Western Europe, Japan, and a few select resource-rich developing countries. It is quite likely that a new round of traditional licensing agreements will be negotiated between U.S. firms and the developing countries as they achieve adequate levels of manufacturing know-how. The latter still are important export markets for U.S. and European-manufactured aircraft equipment.

Future Trends

Replacing conventional licensing will be close collaboration with a foreign partner in an arrangement that provides for some investment by the latter in development costs and other "front-end" costs of a program, a sharing of risks, and an agreed division of manufacturing and marketing tasks. This new approach may involve the granting of licenses but only if coupled with substantial equity participation in the licensee or offset procurement and production agreements or coproduction programs.

Both coproduction programs and offset agreements, in effect, provide for an agreed sharing of the manufacturing tasks in a major program. They thus satisfy, at least in part, the desires of foreign governments and foreign manufacturers for what may be a sizable share of the manufacturing activity, foreign-exchange earnings, and employment while maintaining a significant and specified share for the U.S. licensor. In other words, under such arrangements, someone else can do the sheet-metal work if the U.S. firm can retain the production share that requires the maximum productivity and the investment in heavy machine tools. Given such a division of labor, capital-intensive U.S. production efficiency can offset lower labor costs overseas. The CFM56 engine program between General Electric and SNEC-MA fits this pattern, as does the procurement of General Dynamics's F-16 fighter by four NATO governments.

If collaboration with foreign producers can limit the threat of direct competition by licensees and minimize nationalistic obstacles to the exploitation of foreign markets, there are obvious defensive reasons for preferring it to simple licensing. Other developments, however, are generating increasingly powerful affirmative motives for entering into coproduction programs or acquiring substantial equity interests in foreign ventures. They are the increasing cost and decreasing number of major military and commercial programs and the declining role of military R and D in the development of commercial aircraft and engines. In their military procurement, foreign governments are increasingly turning to multinational programs carried out by consortia of manufacturers in order to share both the costs and risks (especially those of development programs) and the benefits associated with the business generated for their manufacturers and the latter's employees.

The same considerations that motivate governments, namely, the magnitude of the costs and risks associated with new programs, bear even more powerfully on major commercial programs for which no government funding is available (at least in the U.S.) and which nowadays are apt to call for the development of equipment that is entirely new in design, rather than a derivative of military systems. It should be remembered that the first generation of commercial aircraft gas turbines was made up entirely of only slightly modified military power plants whose development and product improvement had been supported by U.S. government funds. Likewise, the Boeing 707 aircraft, one of the first generation of jets, was a derivative of a military aircraft. In recent years, however, there has been notably less opportunity for such adaptation of military systems to commercial use.

Given the need to provide corporate funding for complete development programs in an environment of high interest rates and tight capital markets, it is less and less attractive to companies to act alone, and they have an increasingly strong incentive to form consortia so as to spread the cost and risk. Such an arrangement provides the aircraft company with considerable protection against the effects of political and economic obstacles to exports. It offers a broader financial base for major programs and largely forestalls competition by a licensee with the licensor. Finally, it generates a far stronger incentive than a conventional license agreement to ensure that the technology transfer is effectively accomplished and that it flows in multiple directions. This seems to be the pattern of the future for the aircraft industry.

It is the latter quality of such arrangements that may prove to be a double-edged sword. The dominance of the U.S. aircraft industry in international markets has largely been a function of its superiority in technology. As the incidence of joint ventures and manufacturing consortia among U.S. aircraft firms and foreign firms increases, the effective and complete transfer of highly sophisticted technology, design and engineering know-how, and advanced processes to highly absorptive and receptive environments will be inevitable. Once the foreign partner feels that the arrangement has outlived its usefulness and given its considerably enhanced capabilities, it can become a formidable competitor of the U.S. partner and other segments of the U.S. aircraft industry.

Although activity in the U.S. aircraft industry has declined overall during the past five or six years, it is still a vital part of the U.S. economy. It is basically composed of the major divisions of three giant aerospace corporations—Boeing, McDonnell Douglas, and Lockheed. In addition, it involves a great many other companies that supply parts and components—engine manufacturers such as Pratt and Whitney and General Electric; major subcontractors such as Fairchild, General Dynamics, Grumman, Northrop, and Rockwell International, which build huge sections of aircraft; component makers such as Rohr, Collins, Garrett, and Menasco; and myriad others in various fields that contribute key but smaller pieces of the aircraft.

In the late 1960s, the U.S. aerospace work force numbered about 1.4 million, compared with some 250,000 in Britain and 100,000 in France. At the end of 1975, the industry's employment stood at 938,000, compared to 976,000 a year earlier, and that total is expected to drop to 890,000 by the end of 1976. Total aerospace industry sales ran about $30 billion in the late 1960s, declined to a low point of less than $23 billion in 1971, and have since recovered, in current dollars, to about $27 billion. In constant 1968 dollars, however, industry sales have declined almost steadily since 1968; they stand now at the equivalent of about $18 billion.

As an export product group, only agriculture has had a positive balance of trade greater than aeronautics; it earned over $5 billion in export sales in 1974 and about $3 billion last year. In 1974, two-thirds of the commercial jets and jet engines in worldwide service were of U.S. origin. Despite these impressive figures, however, the world commercial transport market has been declining. Worldwide deliveries of U.S.-built transport aircraft fell from 332 units in 1974 to 282 in 1975. Currently, according to some estimates, the major airlines of the world could handle a 10 percent growth in passenger traffic without buying any new aircraft; and the most optimistic forecasts suggest that there will not be another substantial round of transport aircraft buying until the early 1980s.

The military export market, however, continues to expand, and according to various published sources, it is likely to reach about $12 billion in

1976, the bulk of which will be in government transactions. Northrop at the beginning of the 1976 was reported to have a backlog of 400 undelivered F-SE/F aircraft, all for export, and another 100 orders were expected in the near future.

The following sections briefly survey the competitive positions and capabilities of the major foreign aircraft industries.

The Soviet Union

The Soviet Union's aircraft industry is the second largest in the world. Although it has produced about one-fifth of the aircraft in commercial use, the Soviet industry has exported only about thirty of these planes to non-socialist nations. For about the last two years, it has attempted to create a large export market for its YAK-40 medium-range transport, thus far with limited success.

At least one source in the U.S. industry believes that the Soviet Union is currently building sufficient excess capacity to enable it to support major penetrations of Western aircraft markets.[1] Such claims may be exaggerations or self-serving, for the Soviet record of the last twenty years is a very poor one. Starting with the IL-18s that were returned by Ghana in the late 1950s and continuing through the TU-154s that were returned by Egypt two years ago, Soviet aircraft generally have been unable to compete on the basis of standard airline measures of direct operating cost, availability rates, and maintenance intervals. The beginning of a more significant export potential might be developed if Rolls Royce licenses the RB.211 to the Russians, as is currently under discussion. Even claims to this effect, however, may be an overstatement of Soviet capabilities and absorptive capacity at the present time.

Japan

At the end of World War II, the Japanese aircraft industry was largely destroyed. As part of the peace treaty imposed on Japan, both production and R and D related to aircraft were prohibited until 1952. The rebuilding of the Japanese industry began with repair and overhaul contracts from the U.S. Air Force and the incipient Japanese Air Self-Defense Force. In the mid-1950s, in order to lessen the U.S. defense burden, the Department of Defense sought to have the Japanese assume responsibility for regional antisubmarine warfare and shore patrol. Although U.S. firms would have welcomed the opportunity to export planes to this nation, the Japanese government pleaded balance-of-payments problems. American aircraft manufacturers were therefore encouraged by the Defense Department to license their aircraft designs and technology.

Several models of aircraft have been produced in Japan under coproduction agreements. In these cooperative programs, which usually relate only to the procurement of military aircraft, portions of planes are manufactured in each coproduction partner's facilities. These parts are then mated in any of the aircraft produced under license.

Although the Japanese industry is not a major competitor to U.S. firms, it has in recent years achieved a small volume of export sales, the most notable example being the YS-11, a short-range turboprop civilian transport. The Japanese constitution still prohibits the export of arms, including military aircraft and parts. Total employment in the Japanese industry is approximately 26,000 people.

Western Europe[2]

Currently, the U.S. industry's major foreign competitor is the Western European aircraft industry. The British aircraft industry came out of World War II as a vigorous competitor of the United States and virtually its equal technologically. The French industry, on the other hand, went through a long and arduous process of rebuilding to reach its present level. Gradually, they have come to approximate one another in the value of their production. Now, along with Germany and, to a less extent, Italy, they have started down the hard road of international collaboration discussed earlier.

Currently, the French industry numbers a little over 100,000 workers and the British industry about 200,000, compared to more than 900,000 in the United States. In France, the principal firms are the nationalized aircraft firm, Aerospatiale, with about 40,000 workers; the still private Dassault-Breguet, with 15,000; the major engine firm, SNECMA, with about 12,000 workers; and the smaller but very impressive missile manufacturer, MATRA. In Britain, the main groupings are, in order of size, the Hawker Siddeley Group, British Aircraft Corporation, and Rolls Royce. Together they represent about three-quarters of industry employment.

A recurrent problem of the European aerospace industries has been a low rate of productivity compared to the United States. Social charges to the employer as well as lower investment in capital equipment have created a situation in which, for Europe as a whole, productivity may be about half that of the United States. Differences in wages as well as productivity rates, based both on differing social security practices and on levels of capital investment, are a continuing source of difficulty in harmonizing international programs, both within Europe and with the United States.

Throughout the postwar period the British and French industries have been confronted by the problem of remaining technologically current with the Soviet Union and United States—at least on a selective basis—without the large domestic procurements and R and D budgets that sustain the U.S. and Soviet industries. This situation has led, in turn, to the need for a large export share of total production of tactical aircraft and missiles, in order to support both R and D and capital investment. The French and British industries are inherently export dependent if they are to remain viable in the development and production of a sufficient range of military aircraft and missiles, and this dependence exists, although to a smaller degree, even in the case of major collaborative projects among the European partners.

Case Study: The General Electric-SNECMA JointVenture for the Production of the CFM56 Engine

In 1971, General Electric and France's Société Nationale d'Etudes et de Construction de Moteurs d'Aviation (SNECMA) agreed to jointly develop, manufacture, and market an aircraft engine to be known as the CFM56. At that time, both firms were interested in substantially advancing the state of the art in commercial engines, but neither was willing to undertake such an effort by itself. GE had yet to recover its enormous investment in an earlier aircraft engine—the CF6—and was hesitant to embark on yet another major commercial engine development. SNECMA lacked the commercial experience necessary to give its products credibility with the major privately owned airlines and airframe manufacturers. Under these circumstances, a joint venture had considerable appeal to both firms.

Under the 1971 agreement, a cooperative program was to be established on as nearly an equal basis as possible. GE would provide the main engine controls and "core engine"—the portion where fuel injection and combustion occur. It intended to satisfy these requirements by adapting the F101 core it had developed for the U.S. Air Force's B-1 strategic bomber. SNECMA would furnish the low-pressure engine system, including the fan, engine frame, thrust reversers, gear box, and accessory installation. It would be responsible also for the "systems integration," the mating and fine-tuning of each company's portions of the prototype. Neither firm expected to gain access to the other's designs and technology.

A highly diversified company, GE was one of two large producers of jet engines, the other being the Pratt and Whitney division of United Aircraft Corporation (now known as United Technologies). Although both had achieved a strong position in the production of military engines, Pratt and Whitney clearly dominated the U.S. market for commercial applications. In Western Europe, this market was primarily served by Pratt and Whitney and the British firm Rolls Royce. By successfully promoting its CF6,

CF100, and CJ610 engines, GE had begun to strengthen its position in these markets in recent years. Competition with Pratt and Whitney required that GE continue to upgrade its technology.

SNECMA, of which 80 percent was owned by the French government, 10 percent by Pratt and Whitney, and 10 percent by the general public, was primarily noted for the engines it produced for French military aircraft. In the early 1970s, however, as part of a five-year planning exercise, the French government decided that this firm should play a more substantial role in the world aircraft industry. In order to achieve this end, the French government reportedly offered to provide SNECMA with up to $200 million in loans if it would establish a joint development program with a foreign firm that enjoyed an outstanding reputation in commercializing new products.

For several years, aircraft industry analysts projected that an important potential market might soon develop in short-to-medium-haul flights up to 2200 miles. Due to changes in the regulatory environment and international flight patterns, such flights would ultimately require highly efficient, low-noise, low-polluting engines with between 20,000 and 30,000 pounds of thrust—the so-called "ten-to-fifteen-ton" engines. Virtually no commercial engine existed between the 15,000 pounds-thrust Pratt and Whitney JT8D and GE's 40,000 pounds-thrust CF6. Market analysts predicted that from the late 1970s to the mid-1990s, between 6000 and 8000 of these engines could be sold at about $700,000 apiece. If one were to include the sale of spare parts and accessories, the value of this market might be at least $10 billion. GE analysts felt that capturing as much as one-half of this market might be possible.

The eventual acceptance of the new engines was highly uncertain. They were not expected to be retrofitted onto existing aircraft. Rather, it was believed that ten-ton engines would be utilized to power the next generation of commercial jets, including stretched versions of such planes as the Boeing 727 and 737, and the McDonnell Douglas DC-9; STOL (short takeoff-and-landing) aircraft then in the development phase; and wide-bodied, low-density planes such as the Boeing 7X7 just being conceptualized. Potential customers might also include a 154-seat version of the French Dassault Mercure and a number of other European airplanes planned for the medium-range category. In short, "ten-to-fifteen-ton" engines would not be suitable for any aircraft then in production.

Developing new commercial engines requires considerable investment and acceptance of high risk. Research and development, tooling, and inventory costs for a new engine would be expected to vary between $250 million and $1 billion. Assuming a break-even were ever achieved, the payback period for such an investment could be as long as ten to fifteen years. As a consequence, there were and continue to be few major competitors in the world commercial engine industry.

After the decision of the French government to encourage SNECMA to assume a more active role in international markets, the firm invited GE, Pratt and Whitney, and Rolls Royce to submit independent proposals for joint development of a ten-ton engine. Discussions with the latter were never quite serious because by the early 1970s Rolls Royce was approaching bankruptcy due to its joint development program with SNECMA on the engine for the Anglo-French supersonic transport, the Concorde. In addition, SNECMA felt that partnership with an American firm would lend its expansion efforts greater credibility in international markets.

SNECMA signed conditional agreements with both U.S. companies but left the final decision to the French government. GE was ultimately chosen, perhaps because it was well along in its development of the F101 core, which could be expected to provide a sound basis for the non-French portion of the joint development. Developed for the B-1 bomber, at a cost to U.S. taxpayers of approximately $400 million, and also the recipient of about $200 million from NASA's "quiet engine" program, the F101 core represented the most advanced engine technology in the U.S.[3]

While GE was not a dominant firm in the U.S. industry, it brought many advantages to the liaison with SNECMA. Its performance on the CF6 had earned the company a high reputation for technical competence, product reliability, and adherence to schedules. Although not more respected than the Pratt and Whitney label, the GE brand name on a product symbolized quality and reliability and the French felt it would be a commercial asset. In addition, the very fact that General Electric was not the dominant firm in the U.S. industry provided the French with some reassurance that their partner would not assume a superior attitude in the relationship.

The plan for the joint effort was fairly straightforward. Each company would manufacture its components according to the performance and physical characteristics specified in an overall CFM56 technical requirements document. Although neither firm would divulge its costs or techniques of manufacture to the other, at the end of the development phase they would exchange technical drawings in order to negotiate a division of the sales revenue on each CFM56 engine. Once an engine was slated to be sold in Europe, the GE core would be shipped to the SNECMA plant at Villaroche, France, and assembled with that firm's fan booster and low-pressure system. Similarly, if the sale were to occur in the Western Hemisphere, GE would receive the SNECMA components and perform the final assembly. All sales would be handled through CFM International, a joint sales company to be owned equally by the two firms.

Both governments were expected to respond favorably to these terms. The GE-SNECMA joint venture clearly met the goals of the French government. GE would be a strong partner and would treat SNECMA as an equal. This liaison would help assure successful penetration of the U.S. market for a large volume of French production. The program would have a positive effect on the French balance of payments.

The agreement was also expected to appeal to the U.S. government. U.S. antitrust law, the enormity of the R and D expense, and the decline in both Defense Department-funded R and D and purchases of military aircraft all combined to compel GE in the same direction—toward joint development with a foreign partner. Given the probable decline in military aircraft procurement for years to come and GE's difficulties in gaining a strong foothold in commercial engines, it might be forced to abandon aircraft engine production altogether. Leaving the entire field to Pratt and Whitney would be undesirable from the Defense Department's perspective, because it would no longer be able to play one U.S. company off against the other in order to stimulate development or price competition.

There is another equally compelling reason why the GE-SNECMA joint venture was attractive to the U.S. government. The rise of the European Economic Community had rekindled a sense of European nationalism that had been largely dormant since the end of World War II. A European aircraft-engine consortium, operating behind EEC tariff barriers, would decidedly not be in the U.S. commercial interest. If, for example, SNECMA and Rolls Royce jointly developed their own ten-ton engine, the European market might be closed to U.S. manufacturers, resulting in a loss to U.S. trade flows, national income, and employment. Faced with such a consequence, the U.S. government might have felt compelled to increase its funding of commercial engine developments for years to come.

GE's request for a provisional license to export technical data was approved in December 1971. However, in October 1972, acting on the advice of the White House and the Departments of Commerce, State, and Defense, the State Department's Office of Munitions Control rejected a more detailed version of this request and thus prevented the joint development program from proceeding. The grounds cited were reasons of "national security and foreign policy." Although this information was not clarifying by itself, it soon became apparent that a principal objection centered on GE's plan to export a version of the F101 core. The provisional approval specifically prohibited the transfer of any hardware containing technology more advanced than that of the CF6.

The U.S. government officials who reviewed the licensing request were alarmed by what they understood to be happening. Since much of the technology of the CFM56 core had originally been developed for the B-1 (military bomber) application, many analysts considered its protection vital to the national security. Given that the Communist Party was well represented among the ranks of French workers, advocates of this concern might have found SNECMA's role in the systems integration particularly threatening. It was felt that during the link-up and fine-tuning of the prototype, knowledge of the thermodynamic considerations and management techniques that went into the GE development would be absorbed by the SNECMA technicians and engineers. The transfer of such information would be tantamount to a transfer of technology, that is, design methods, design rationale, and design data. These data would be much more extensive than that which might become available by a mere exchange of drawings.

When the licensing request was vetoed, many people in the U.S. government argued an alternative solution to GE's situation. The Air Force suggested funding the development of a ten-ton engine for use in an advanced medium short takeoff-and-landing transport. This engine could later be adapted for commercial use and, assuming that GE would be awarded the contract, it would no longer need a foreign partner to further its position in the commercial market. The transfer of technology overseas could thus be avoided. Congress, however, refused to fund the project.

According to journalistic sources, the French government was displeased by the U.S. government action. President Pompidou personally sent President Nixon a note expressing his concern about the matter. The French government also reportedly threatened to retaliate by vetoing several proposed agreements between U.S. and French firms. If in fact such threats were made, they indicated more bluster than serious intention, for the French had pending at that time several highly sensitive deals that required U.S. government approval.

After it was denied the original licensing request, GE let some time elapse before submitting a revised proposal. Under the circumstances, the company may have felt that the joint program could not be revamped so as to become acceptable to U.S. regulatory authorities. While GE pondered what additional steps it could take, French interests sought to revive the idea of a European consortium, possibly with Rolls Royce and Fiat as partners. After several months, DGA, Inc., a Washington-based firm representing SNECMA and other European corporate interests, began to make inquiries to the relevant federal agencies. It determined that an acceptable restructuring of the agreement was possible.

With the proposed Air Force-funded program quashed and the Europeans considering the formation of a consortium without U.S. participation, federal officials may have begun to view the proposed GE-SNECMA joint venture in a more favorable light. Certainly, if a European consortium were to develop a commercial ten-ton engine ahead of U.S. manufacturers, U.S. firms and the economy would have much to lose. In order to meet the competition, U.S. jet engine manufacturers might then have had to undertake enormous R and D programs or license the technology from abroad. Perhaps for these reasons, DGA found that U.S. government officials were prepared to consider alternative proposals.

The second round of deliberations concluded that national security interests could be protected with some modifications in the terms of the proposed license. Those portions of the B-1 bomber core-engine technology deemed vital to strategic concerns could be safeguarded by lowering some of the performance parameters of the planned engine. In most cases, these characteristics could be adjusted to conform to those of the CF6-50C, a version of the GE engine already in commercial production. As a result, the technology of the B-1 and that of the new core would differ somewhat; for example, the engines would operate at different temperatures and have different bypass ratios. In order to further placate U.S. government regulatory authorities, the systems integration would occur without SNECMA participation. SNECMA personnel would be completely barred from attendance when the two firms' components were initially linked up and their performance analyzed and balanced.

The revised proposal postponed the actual export of the hardware by a year. This delay was designed to enable SNECMA to receive a ten-ton core no sooner than it probably would have become available from an independent European consortium. The schedule for the delivery of the core drawings also was stretched out.

In addition to the described differences between the two licensing requests, more generous provisions were made for reimbursing the U.S. government for its earlier outlays of R and D funds. Although token federal recoupment of R and D had occurred on several other engines, the CFM56 would be the first development effort in which significant repayment could occur. GE agreed to pay the Department of Defense $20,000 on each CFM56 sold. A high level of sales may produce tens of millions of dollars in royalty payments.

In May 1973, Presidents Nixon and Pompidou met in Reykjavik, Iceland, and formalized the U.S.-French understanding on the GE-SNECMA relationship. According to Edward E. David, Jr., then the science advisor to the American president, one aspect of this agreement related to a tariff concession sought by the U.S.[4] Although U.S. aircraft products could be purchased in the Common Market on a duty-free basis at that time, the U.S. charged a 5 percent tariff on U.S. imports of similar items. After the licensing request had been vetoed, the French intimated that they might petition for the imposition of a countervailing tariff. At Reykjavik, they agreed to refrain from such action for the duration of the GE-SNECMA agreement. Thus, the ultimate disposition on the licensing request was based at least in part on demonstrably commercial considerations.

Since the approval of the revised licensing request, the joint venture has proceeded according to schedule. On 20 June 1974, the first complete CFM56 engine was tested in Evendale, Ohio, and the resulting performance exceeded announced expectations. Test data relating to the French portions of the engine were relayed via satellite to Villaroche. Following the hardware export from the U.S., a second prototype was tested in France. SNECMA engineers were given access only to test data relating their portions of the engine. The two firms were expected to complete their exchange of manufacturing drawings by the middle of 1976, the scheduled date of the CFM56 certification.

Pratt and Whitney and its European partners are planning to proceed with the development of the JT1OD. Unlike the CFM56, which will have an initial thrust of 22,000 pounds with growth potential to 27,500 pounds, the Pratt and Whitney engine is designed to have an initial 25,000-pound thrust capability with potential growth in later versions to 30,000 pounds. Pratt and Whitney has announced the testing of a demonstrator engine, but the U.S. firm and its foreign partners have not yet committed themselves to a full development program.

The relative position of Pratt and Whitney, GE, and SNECMA in aircraft engine sales is reflected in the following statistics. In 1974, Pratt and Whitney sold more than $3.3 billion worth of military and commercial engines. Comparable figures for GE and SNECMA were $1.34 billion and $318 million, respectively.

As of mid-1976, this much was certain: by pooling their resources, GE and SNECMA have been able to undertake a development effort that neither could have pursued on its own. The characteristics of the engine they developed are now being studied by airframe manufacturers to determine how designs of their future products might be modified to optimize the engine's performance. In large measure, the development of the next generation of commercial jetliners will be an iterative process, with airframe and engine manufacturers adjusting their products to accommodate breakthroughs in each other's technology.

Case Study: The General Dynamics F-16 Coproduction Program with a European Consortium[a]

On 7 June 1975, the governments of Belgium, Denmark, Norway, and the Netherlands selected the General Dynamics F-16 as the replacement aircraft for their aging fighter fleets. In order to promote greater standardization of equipment within the North Atlantic Treaty Organization (NATO), the four nations had earlier agreed to procure the same type of aircraft. (Other members of NATO did not participate in this agreement.) The F-16 was chosen after an intensive international sales competition that pitted this plane against the Northrop F-17, the improved French Dassault Breguet

---

[a]Because many aspects of the F-16 coproduction program are classified, this case study was written solely on the basis of published sources.

Mirage F-1, and Sweden's Saab-Scandia Viggen fighter. Widely heralded as the arms contract of the century, the selection of this low-cost supersonic airplane will lead at least to $2 billion in sales for the U.S. company. According to conservative estimates, sales of this aircraft may exceed $15 billion over the next twenty years.[5]

As part of the intergovernmental understanding related to this purchase, much of the cost of the consortium's procurement will be subject to coproduction or offset agreements.[6] These offsets will occur on a 10/40/15 percent basis; that is, firms in the four Western European nations will produce a minimum of 10 percent of the value of the F-16s to be procured by the U.S. (currently a minimum of 650), 40 percent of the value of the plans to be purchased by the Belgian, Danish, Norwegian, and Dutch air forces (at least 306), and 15 percent of the value of the sales of the aircraft to all other nations. In order to facilitate direct offsets, General Dynamics and its major subcontractors will transfer substantial portions of the technology required to produce F-16 components in Europe. According to the official policy of the U.S. government, no profits may be earned on the licensing agreements eatablished for this purpose. U.S. national security considerations require that the transfers be restricted, in the near term, to exclude such items as the aircraft's electronic countermeasure devices (ECMs) and the most strategically sensitive parts of its Pratt and Whitney F100-PW-100 afterburning turbofan engine.

One of the most unique aspects of this program is its coproduction schedule. In previous defense-related agreements, delivery to the U.S. weapons inventory has occurred several years ahead of foreign sales. In the F-16 program, full-scale U.S. production will commence in 1979 and will be followed a year later by European mass production. Thus, the technology transferred to firms in the consortium countries will not be fully proven by mass production before it is implanted. In this and other regards, the transfers of production technology will be truly at the state-of-the-art level. The aircraft itself is said to be a revolutionary step forward.

The sales price for each plane has been fixed at a minimum of $5.16 million and a maximum of $6.09 million. The lower figure includes $2.77 million for the airframe, $1.2 million for the engine, $670,000 for U.S. Department of Defense research and development recoupment, $370,000 for the target detection systems, and $150,000 for U.S. government-supplied equipment. The higher figure incorporates an anticipated level of inflation.[7]

General Dynamics has been less severely affected than other firms in the industry by the recent decline in government defense procurements and civilian aircraft sales. The company has continued to be awarded a number of extremely large U.S. government contracts. Besides being the prime contractor for the F-16, General Dynamics produces the F-111 (the world's first swing-wing aircraft), Trident ballistic missile submarines, sea-launched Tomahawk cruise missiles, and the Atlas and Centaur rockets used by the U.S. space program. Among other items, the company also manufactures fuselages for the McDonnell Douglas DC-10 wide-bodied jet transport and 125,000-cubic-meter liquefied natural gas tankers.

In 1975, General Dynamics earned $84.5 million on sales of approximately $2.2 billion. Aerospace sales contributed 30.4 percent of these revenues and 37.8 percent of corporate net income. Although the company operates plants and offices in ten countries, it earned less than 10 percent of its 1975 revenues from foreign sales. The principal beneficiary of General Dynamics's F-16 contract is its plant in Fort Worth, Texas; the size of its work force is expected to double as a result.

The F-16 is a multimission aircraft combining advanced technology with low cost. It has twice the maneuverability of current fighters and three times the combat radius in an air superiority role.[8] Because it is relatively small (only 8 feet long, 16.5 feet high, and with a wingspan of 31 feet) and lightweight (about 22,500 pounds), the F-16 is expected to be a difficult target while it is flying at twice the speed of sound and at an altitude of more than 60,000 feet.

According to the company's press releases, no other fighter, planned or in operation, makes such extensive use of advanced technology. The F-16 incorporates features never before combined in a single aircraft: forebody strakes—for controlled vortex lift; wing-body blending—for greater body lift and reduced drag; fly-by-wire electronic flight controls; a variable wing camber with automatic leading edge maneuvering flaps; side-stick pilot's control; and a high-acceleration-high-visibility cockpit. The technologies incorporated in this plane were selected and integrated so as to decrease the vehicle's weight, cost, difficulty of manufacture, and fuel consumption. While the F-16's combat capability has been described as far superior to previous fighters, it requires only about one-half the investment, fuel, and manpower. Its armament includes Sidewinder missiles, 500-pound bombs, and an internally mounted Vulcan 20-mm machine gun.

By September 1975, General Dynamics had placed orders with fifty-four U.S. firms for various F-16 components and systems. Subcontractors on the project include General Electric, United Technologies, Westinghouse, Singer Kearfott, Bendix, Hughes Aircraft, Airesearch Manufacturing, TRW Systems, Teledyne Electronics, Sperry Rand, Sylvania Electronic Products, and Brunswick Corporation. Each of these companies is required to arrange for offset production in at least one of the four consortium countries. Ultimately, many of the components of the F-16 will be produced both in the United States and the consortium nations. Thus production will involve several hundred corporations in the two areas. Although final authority for these arrangements resides with the Department of Defense, General Dynamics is responsible for their coordination and for verification that each nation receives at least its promised share of offset production.

The U.S. company is said to be employing the following criteria in its selection of coproduction partners. It is seeking to utilize existing industrial capabilities, to contribute to the technological upgrading of the four NATO nations' industrial base, to support local employment for extended production runs, to provide the basis for in-country support of the aircraft during its life cycle, and to provide retention of program costs in each country.[9]

The minimum levels of offset production were established through intergovernmental negotiation. The announced plan is to return approximately $750 million worth of business to Belgium (92.99 percent of its F-16 purchase), $328 million to Denmark (80 percent of the cost of its procurement), $406 million to Norway (80 percent), and $697 million to the Netherlands (98.3 percent).[10] The figures are based on the expectation that 2000 planes would be produced. In fact, the potential world market for this aircraft has been estimated to be 4880 planes.[11] Under the terms of the U.S. consortium agreement, the overall 88 percent offset will be achieved when 1500 F-16s are purchased. After more than 2000 planes have been sold, the four countries will obtain more than a 100 percent offset.[12] It should also be noted that when 850 planes are sold outside the U.S., there will be a $3 billion positive impact on the U.S. balance of payments.[13]

Two lines have been set up in Europe for final assembly of the consortium's F-16s—one in Belgium and one in the Netherlands. The Belgian assembly operations will be carried out by Fairey and SABCA and the Dutch by VFW Fokker. These firms will also participate in the following aspects of the coproduction program. Fairey will manufacture the aft fuselage and vertical fin box and will mate the airframe sections. SABCA will build the wing boxes and VFW Fokker the leading edge flaps and trailing edge flaperons.

Although the F100 engine coproduction program calls for Fabrique Nationale, a Belgium company, to handle the final assembly and testing of the engine, the production of various components has been subcontracted to firms in each of the four countries. In addition to its other role, Fabrique Nationale will build the inlet/fan and core engine modules. The Danish firm Motorfabriken Bukh will manufacture the gearbox and one of the five engine modules. Kongsberg Vapenfabrikk of Norway will construct the fan drive turbine section, and N.K. Phillips Machinefabriek will build the augmentor and exhaust nozzle module. Belgium has been assigned the largest share of this work—about $190.2 million worth of business, resulting in the temporary creation of at least 3000 jobs. The estimated dollar volume and job impact in the other nations is as follows: the Netherlands, $149.5 million and 1400 to 1600 jobs; Norway, about 500 jobs; Denmark, $13.3

million and up to 1000 jobs. These levels of employment will taper off after 1982, as F-16 production diminishes.[14]

In countries in which adequate levels of direct offset have been difficult to arrange, the U.S. government has sought to encourage indirect offsets. According to speculation in various popular journals, the recent $30 million U.S. Department of Defense purchase of Belgian tank machine guns was arranged partially for this reason.[15] As a result of this procurement, similar machine guns will no longer be bought from the Maremont Corporation plant in Maine. As a consequence, approximately 500 production workers will be laid off. According to a General Accounting Office report, the Belgian weapons cost more than twice as much as those produced by Maremont. On the other hand, the Defense Department claims that the Belgian machine gun is clearly superior.

Several potential problems with the coproduction arrangements have become evident. VFW Fokker has openly considered shifting some of its production for the F-16 program to its subsidiary in West Germany. Westinghouse successfully underbid Hughes Aircraft on a contract to supply the U.S. airborne radar equipment. As a consequence, the anticipated level of European electronics offset production has been somewhat reduced. The U.S. General Accounting Office has expressed concern about the inadequacy of the auditing procedures for the coproduction contracts. Finally, differential rates of inflation in the consortium nations may further complicate the accounting problems.

Sales of the F-16 outside the U.S. and the other four nations are subject to the approval of the U.S. Departments of Defense and State. The governments of Israel and South Korea have already expressed an interest in this aircraft and have suggested establishing offset production requirements. (Israel was recently allowed to make a direct purchase with no provision for offset.) Whether the Defense Security Assistance Agency will approve requests for additional offsets is unknown, although, as noted before, sales outside the five nations will have an extremely positive impact on the U.S. balance of payments.

The consequences of the F-16 coproduction program are overwhelmingly favorable to the U.S. The only potentially negative aspect of this program relates to the effect of the transfers on the international commercial environment. Should the transfers greatly strengthen European aircraft component manufacturers, specific U.S. companies may suffer a loss of business. There are several strong arguments in favor of discounting the importance of this problem. First, the magnitude of the sales gained as a result of the coproduction program cannot conceivably be outweighed by the potential loss for the U.S. economy of a few marginal sales. Second, U.S. firms will retain a competitive advantage because they performed their own R and D. It also seems reasonable to assume that U.S. companies will upgrade their technology in areas where they suspect they may become vulnerable to competition.

The F-16 coproduction program is the most extensive ever attempted. Besides furthering the goals of the U.S. and the consortium governments, it will have a positive impact on the economy and defense capabilities of every participant nation. It should be understood that coproduction was not solely an act of goodwill; it was a prerequisite to the sale established by the four NATO governments.

Case Study: The Piper Aircraft Corporation Licensing Agreement with Empresa Brasileira de Aeronautica, S.A.

In 1974, Brazil represented the largest single export market, outranking both Canada and Germany, for U.S. light aircraft (general aviation) manufacturers. U.S. manufacturers delivered 726 planes to Brazil in that year at a cost of $600 million. Severely pressed by this time with foreign-exchange constraints and confident of its technical capabilities and sufficient internal market demand, Brazilian development authorities felt it was an appropriate moment for the state-owned aircraft enterprise, Embraer (Empresa Brasileira de Aeronautica, S.A.) to begin a manufacturing program of light aircraft, single- or twin-engine, with the support of and in close cooperation with a foreign aircraft manufacturer.

Embraer was created in the late 1960s for the express purpose of promoting the development of the local aircraft industry. After six years of operation, it had three lines of aircraft in production: the "Bandeirante" EMB-110, a derivative of the French Nord 262; the "Ipanema" EMB-201, a single-engine crop duster designed by Embraer; and the "Xavante" EMB-326GB, a jet trainer and ground attack aircraft produced under license from Aeronautica Macchi S.p.A., an Italian corporation. As of late 1974, Embraer employed 3500 people and had a total capitalization of about $20 million.

In 1974, Brazil sent a mission to the major U.S. small aircraft producers to solicit proposals on an agreement for production by Embraer in Brazil of U.S. planes. Embraer approached Piper, Beech, and Cessna, the last of which held more than 60 percent of the Brazilian market in 1974.

According to Embraer, all three firms were fully apprised of the rules of the game; that is, the Brazilians made explicit their intent to develop their own technical, managerial, manufacturing, and marketing capabilities in small aircraft production and to reserve exclusively the domestic market thereafter for Brazilian-produced aircraft. The second goal, it was explained, was not so much an intent to create a protected industry but an effort to realize foreign-exchange savings. Implicit in these rules was the eventual outcome that only the foreign firm prepared to enter into an agreement with Embraer would be permitted continued participation in the large Brazilian market.

In the early phases of the negotiations, competition among the three U.S. firms was spirited, especially between Cessna and Piper. Initially, Embraer had a marginal preference for Cessna because it enjoyed wide recognition and confidence within the country, due to its large market share and effective distributorship system. Beech dropped out as a serious contender quite early, taking the position that if Brazil wanted its aircraft, it would have to import them from its U.S. facilities.

From all appearances, Cessna initially entered negotiations in earnest with a preparedness to release technology and managerial control to Embraer for production of its aircraft. Its ultimate position, however, was not unlike that of Beech's. Evidence of Cessna's true intent was its adamant refusal to grant Embraer authority to make modifications it deemed appropriate in the Cessna aircraft models the company chose to manufacture. To anyone familiar with Brazil's goals in developing its own aircraft industry, such authority would be expected to be a core feature of the kind of industrial cooperation agreement sought by Embraer. Cessna's attitude on this issue suggested a fear that the quality or performance standards of its aircraft would suffer if it agreed to this term—a suggestion that was not lost on the Brazilians and was highly offensive to Embraer's sensibilities on this subject and to its estimation of its own capabilities. A second difference that arose in negotiations between Cessna and Embraer concerned royalty payments. Embraer wanted no royalty obligation for manufacturing know-how acquired from the foreign partner, and Cessna felt it was a legitimate term to the agreement.

Undoubtedly, several more subtle differences arose between Cessna and Embraer during the negotiations, but the outcome was that Piper was selected. As was portended by Embraer officials prior to entering into negotiations, U.S. exports of small aircraft to Brazil have plummeted. Cessna, which in 1973 sold more than 400 aircraft in the Brazilian market, sold only 5 in 1976. This sales plunge is the result of a 50 percent tax (raised from 7 percent) imposed in 1975 on imported planes of this category and the requirement of the Brazilian government that importers make a one-year interest-free deposit covering the full price of manufactured goods bought abroad. In addition, Brazil's law of similars, which has been in existence since the 1890s but of limited application until recently, has severely inhibited aircraft exports to Brazil. The law stipulates that once an item is produced in "sufficient quality and quantity" in Brazil and registered with the government as a product similar to its imported counterpart, it will be protected from imports.

The industrial cooperation program with Piper is based upon two agreements—one for single-engine aircraft and one for twin-engine airplanes. Each will be operative at least through mid-1979. Under the terms of the agreements, Embraer may select any Piper model it desires for local production. Thus far, the following models have been chosen: three models of Cherokee aircraft and the Lance, Seneca, and Chieftain.

Piper is responsible for providing the necessary assembly and parts manufacturing know-how as well as for assisting in such areas as quality control, materials handling, and manufacturing. Piper has an option to use its international distribution system for aircraft that may be exported from Brazil. The U.S. firm's compensation is primarily a percentage return on the components it ships to Embraer. As the licensee progressively substitutes local content for these imports, the returns will diminish. However, even at 100 percent production in Brazil, Piper still will be paid a fee for service in support of those aircraft. With the exception of those items that cannot be economically produced in Brazil, local substitution is expected to proceed smoothly.

At the present time, the Piper program is basically a licensing agreement; but in the medium and long term, it could provide for the cooperative development of new aircraft. The agreement specifically permits Embraer to fabricate Piper aircraft for sale in the domestic market and, on occasion, to produce jointly with the U.S. company for foreign market sales; replace on a gradual scale Piper-supplied components with Embraer-fabricated products; initiate joint programs to share development and production of a new aircraft aimed at domestic or foreign markets; and market one another's products through individual distribution networks.

Production capability for the Piper models is being transferred to Embraer in three phases. During Phase I, completed structures such as fuselage, empennage, and wings are shipped to Embraer for final assembly and installation of all systems and components. Phase I was completed for the single-engine models in six months and Seneca is now also in Phase II. During Phase II, Embraer receives structured subassemblies for mating in jigs, in addition to the functions achieved in Phase I. By the Phase III, Piper will be shipping all component parts for assembly by Embraer and in three subphases will: (1) begin replacement of Piper-supplied parts by Brazilian-made equivalents, including interiors and 50 percent of both fiberglass and acrylics; (2) complete replacement of all remaining fiberglass and acrylics and produce all harnesses; and (3) produce the aircraft completely with Brazilian-manufactured parts and components, with the exception of those that cannot be economically produced in Brazil.

Upon completion of Phase III-3, Embraer projects that from 66 percent to 70 percent of the Piper product will be of Brazilian origin based on U.S. price. Single-engine models will reach Phase III-3 by mid-1977, the Seneca by the end of 1977 and the Navajo by mid-1978. Subcontracting is a central feature of the Embraer-Piper production program as well as Embraer's other aircraft programs and was instrumental in allowing the company to begin these projects so quickly. Subcontracting is also vital to the development of other aviation-related industries around Embraer. Over fifty Brazilian national firms participate in the subcontracting network for the country's aerospace industry.

The development of the Brazilian aircraft industry has been shaped by three ten-year plans spanning the period from 1950 to 1980. A government-funded organization known as Centro Technio Aerospacial (CTA) has been given responsibility for directing this program. CTA performs many of the functions handled in the U.S. by NASA and the Federal Aviation Administration. In the first ten-year period, CTA sought to establish a teaching and training program to develop a support structure for the aviation industry. The second period required the establishment of technically strong local manufacturers. The third period was characterized as one of increasing the sophistication of locally produced power plants, avionics, and aircraft systems that will go into Embraer products. CTA has aided Embraer by helping it to modify its Bandeirante designs.

The primary means employed to develop these supporting industries is a series of partnerships with foreign firms. An effort is made first to make each industry financially capable and then to upgrade its technical competence to aeronautical standards. To assist and accelerate the latter process, one foreign firm is normally selected for each industry.

In developing its aeronautics industry, Brazil's specific areas of interest concerning foreign partnerships include both reciprocating and turbine engines, avionics, hydraulics, instrumentation, and raw materials such as aluminum and steel. CTA has begun earnest discussions with a number of foreign firms for the following forms of assistance. A formal agreement is expected to be signed soon with Lycoming whereby it will help in the development of Brazilian components to be produced by individual vendors. These components include raw materials, forgings, castings, machine parts, and accessories. Teledyne Continental Motors has been approached by CTA for similar assistance. In the field of turbine engines, talks are underway with Pratt and Whitney Aircraft of Canada, Rolls Royce, Lycoming, Garrett, and, to some extent, with General Electric. CTA has been working with Collins, Bendix, King, Narco, and the German firm Becker for the possible production of avionic electronic components in Brazil. A working relationship with the French has already been established for the manufacture of hydraulic components, including landing gears, and other French firms have been engaged in negotiations for production of light alloys and resistance steels. And for the production of flight and power plant instruments, CTA is holding discussions with three U.S. and British companies.

Although the future of Brazil's aerospace industry looks promising, there are certain weaknesses that only experience and competition will overcome. Embraer is highly engineering-intensive and therefore will experience difficulty in keeping costs down. Regardless of how well engineered an aircraft is, unless it is price competitive in the international market, it will not be a successful seller. Price is especially important if Embraer expects to sell 50 percent of its EMB-12X line in the export market against such fierce competition as the Beech 100/200 and the Swearingen Merlin. Retail base prices of Embraer's Piper models average 27 percent above those charged by Piper in the U.S. Increases range from 17 percent for the Chieftain to 40 percent for the Arrow.

Embraer-produced aircraft also at present suffer from low-quality interiors. The materials now being utilized will not hold up under hard use. Piper officials have indicated that they will do the completions in the U.S. should they choose to market the new pressurized aircraft through its domestic and worldwide distribution system. Another deficiency in Embraer is its lack of effective marketing tactics. Its program has been criticized by Brazilian dealers experienced in high-pressure international markets for being insufficiently aggressive.

While the rest of the South American market for light aircraft appears to be sewed up by assembly programs of Cessna and Piper in Argentina and in Colombia for the Andean Pact countries of Venezuela, Ecuador, Peru, Bolivia, and Chile, the African market is still quite accessible. Brazilian marketing efforts can be expected to aim first in that direction. If the Brazilians have aspirations of marketing their EMB-12X models in the U.S. market, they must first drop some of their own high trade barriers. The 123 and 120 could not possibly compete with Beech and Swearingen on an equal footing with a similar 50 percent import tax imposed on the airframes coming into the country.

Brazil is not unique among the more industrialized of the developing countries in its desire to develop an indigenous aircraft industry. It is perhaps unique, though, in its professional approach to achieving that end. Its strategy of effectively closing entry to its markets for all but the foreign firm prepared to share front-end technology, to impart sophisticated design and engineering capabilities, and to instruct Brazilian nationals in managerial skills has been extremely successful. Foreign firms, facing narrowing opportunities to earn returns in foreign markets, have been extremely eager to meet these conditions. Brazil, a state capitalist economy, has the means to accumulate large sums of capital and therefore can exercise considerable leverage in negotiating with foreign firms.

In time, the sustained enterprise-to-enterprise relationship that characterizes Brazil's effort to develop its aerospace industry will result in the implantation of internationally competitive design, engineering, and production capabilities. While the U.S. firms that have taken advantage of the opportunities offered by Brazil will benefit, other U.S. firms in the industry can expect increased competition in third-country markets from Brazilian aircraft in the future.

Notes

1. Aerospace Research Center. *The Challenge of Foreign Competition.* Washington, D.C., November 1975, p. 2.

2. This section draws heavily on a paper prepared by John Hoagland, entitled "American and European Industry Relations in Military Exports to the Less Developed Countries," for a Conference on the Military Build-up in the Non-Industrial States, sponsored by the Fletcher School of Law and Diplomacy at Tufts University, 1976.

3. Following the selection of GE by the French government, Pratt and Whitney decided to proceed with the development of its own ten-ton engine, the JTTOD. This effort was to involve the participation of Rolls Royce, Fiat, and the German consortium MTU but was not designed to be a collaboration among equals. Rather, Pratt and Whitney wanted R and D funds from the others, and the foreign companies would serve primarily as investors and subcontractors.

4. Edward E. David, Jr. "Technology Export and National Goals." *Research Management* (January 1974), p. 14.

5. "The F-16 and How It Won Europe." *The New York Times*, 27 July 1975 (Week in Review), p. 1.

6. According to Erwin J. Bulban, "U.S. Sets Fighter Sales Price to NATO," *Aviation Week*, 24 February 1975, p. 14, in early 1975 General Dynamics offered the NATO consortium countries a minimum of an 88 percent offset.

7. Ibid.

8. *General Dynamics News*, 23 September 1975, p. 1.

9. "European F-16 Suppliers Detailed." *Aviation Week*, 27 October 1975, p. 81.

10. Ibid.

11. "F-16 European Co-Production Programme." *Interavia*, August 1975, p. 872.

12. Ibid.

13. "Washington Roundup." *Aviation Week*, 15 September 1975, p. 13.

14. "U.S. Clears F100 Technology Transfer." *Aviation Week*, 24 February 1975, pp. 16-17.

15. "Guns from Belgium." *Business Week*, 12 April 1976, p. 40.

3

The Automotive Industry

Sector Overview

The U.S. automotive industry has been undergoing profound changes in the past twenty-five years. In the 1950s, the U.S. supplied three-fourths of world demand for passenger cars and trucks. By the 1970s, this share had dropped to a third of world demand. As markets developed abroad, U.S. industry was forced to move production facilities abroad in order to avoid losing these markets. These outward movements have entailed mounting capital investments for the proliferation of production facilities. Foreign manufacturers in Japan and Western Europe have begun to make serious inroads into the U.S. market, challenging U.S. producers and cutting into earnings sensitive to maintaining volume production and utilization of installed capacities. These difficulties have been further compounded by enlarged demands for R and D expenditures to meet U.S. government and consumer demands for vehicle safety, fuel economy, and pollution-free engines.

The combined impact of changed market environments in newly industrializing countries, tough worldwide competition from foreign automotive firms, and a general decline in production levels at home has made it imperative that even financially strong U.S. firms such as General Motors adjust their corporate philosophies and production practices worldwide. A newly negotiated GM-Polmot agreement represents the forward edge of corporate adjustment to a changing world. The transition from 100 percent equity and full managerial control that characterized GM overseas involvement until a few years ago to the release of technology to a noncontrolled affiliate has had important implications in terms of the implicit commodity flows of capital, earnings, and management resources; the numbers of U.S. production workers; and the long-term competitiveness of U.S. industry, its position in U.S. and world markets, and the implicit impact on U.S. employment, income, and foreign-exchange earnings.

Automotive Technology-Transfer Requirements

Automotive technology consists of vehicle designs, production techniques to fabricate components, and managerial systems to control the quality and cost of production. Cars and trucks consist of a power train (engine, transmission, drive shaft, and axles), body and chassis, and a wide variety of "hang-on" parts (tires, batteries, exhausts, radiators, upholstery, etc.).

Automotive parts are manufactured from hundreds of different kinds of iron and steel and other industrial metals and materials, including rubber, plastic, and glass. Mass production of standardized components and parts demands a rigid uniformity in materials specifications and manufacturing tolerances. A small passenger car averages 2500 major parts and assemblies, or 20,000 parts if every nut and bolt is counted separately. A standard diesel engine for trucks consists of 750 parts provided by about 200 different plants. About 15,000 separate machining and treatment processes are required to run steel shapes, forgings, and castings into finished engine components, such as pistons and engine blocks.

The technical documentation in technology transfer consists of blueprints for plant layouts, process sheets detailing fabrication steps and equipment requirements, quality control and testing procedures, and materials specifications and manufacturing standards. For the manufacture of a diesel engine alone, eight to ten volumes (3000 to 4000 pages) containing materials standards and manufacturing specifications are required. There are approximately 145 technical specifications, engineering information items, testing methods, and engine-rebuilt standards; 67 special manufacturing methods; 439 materials standards; 240 process standards; and 25 salvage procedure standards for rejected parts.

Production of passenger cars in North America, Europe, and Japan is high volume and automated and requires relatively large investment in plant and equipment for vehicle assembly and the thousands of components and parts required for each vehicle. For example, in Brazil, where 100 percent of the vehicle is manufactured locally, 400 firms supply the parts that represent 40 percent of vehicle content.

There are important economies of scale, particularly in the manufacturing of power train and in body elements. Mass production of standardized, quality-controlled components requires hundreds of different types of industrial materials, including iron, steel, nonferrous metals, rubber, plastic, glass, and textiles. Adequate lead times of three to eight years are required to develop the basic materials and processing industries (particularly metal working: castings, forgings, stampings, and machining), and the quality-controlled, production-scheduled systems that go with them.

The transmission of technical knowledge related to industrial techniques requires a high caliber of engineering and technical personnel at both the dispensing and receiving ends. Indispensable to efficient production operations are management systems to plan and install the production system, to schedule the flow of materials and output, to set and control materials and output standards, and to set and maintain functional standards of inputs and outputs at all stages of the production operation.

Trends in World Automotive Industry[1]

World production of automotive vehicles (passenger cars, trucks, and buses) totaled 35 million in 1974, down from 39 million in 1973. Until 1974, production had increased at about 10 percent annually. Production is concentrated mainly in the automotive manufacturing centers of North America, Europe, and Japan and is carried out by large firms—General Motors and Ford account for over a third of world production, less than 10 other firms manufacture another third, and over 300 smaller companies (mainly truck and bus manufacturers) produce the remaining third.

There has been a marked shift toward internationalization of the industry. The share of total world production held by U.S. manufacturers dropped from 76 percent in 1950 to 48 percent in 1960 (as Europeans took over 35 percent of the market) and to 29 percent in 1974, when Japan's share rose to 19 percent. This spectacular shift has forced U.S. firms operating over 200 assembly and manufacturing plants located in various parts of the world to develop and operate complex managerial and technology transfer systems. In 1973, Ford's investment in foreign plants (over $500 million) exceeded new investment in U.S. facilities for the first time.

U.S. firms have also been losing their competitive position in the home market. Compacts from Europe and Japan provided the entering wedge. Imports now account for close to 15 percent of the U.S. market, up from 6 percent in 1964. Foreign imports have made these incursions into the U.S. market despite an almost 50 percent devaluation against the German mark and almost 30 percent devaluation against the Japanese yen. Delayed response by U.S. firms to design adjustments (fuel economy, mainly) has been an important factor. Japanese and German manufacturers dominate the low-horsepower-engine field, for example, motorcycles and snowmobiles. U.S. exports of cars, trucks, and buses, which numbered 661,000 in 1973, are dwarfed by U.S. imports—totaling, in that year, four times that figure.

In the other metal-working industries, U.S. firms now face intensified import competition in the standard ranges of ball bearings, roller bearings, valve fittings, and other metal fasteners. The U.S. industry has managed, however, to expand its exports in the more sophisticated range of mechanical power (transmissions, clutches, axels, and coupling chains), valves and pipe fittings (to meet new requirements for corrosion, pressure, and temperature changes), and antifriction bearings.

Until recent years, when public pressures forced expenditures on safety and pollution-free engines, the pace of technical innovation in the U.S. industry generally has been short of spectacular. Foreign manufacturers have often been in the forefront of change, as evidenced by innovations such as front-wheel drive, disc brakes, radial tires, articulated suspension, and the

Wankel engine. R and D expenditures by the U.S. automotive industry, which have been averaging under 3 percent of sales, were devoted largely to stylistic changes rather than to innovation aimed at improved vehicle performance characteristics.

Investments in automotive plants in developing countries have proliferated in response to national industrialization policies based upon protection and progressive import substitution. U.S. vehicle manufacturers by and large have preferred majority ownership and control of their manufacturing affiliates as the best assurance of obtaining the latitude and flexibility to maximize global profit. From the commercial viewpoint, it gave them the managerial control to protect the product standard and trademark and to link aggressive marketing and financial controls to production operations.

Local content laws required foreign firms to manufacture a progressive number of components and parts, starting with easy hang-on items (tires, batteries, electrical wiring, upholstery, etc.) and followed by more complex auxilliary parts (brakes, clutches, exhaust pipes, radiators, vehicle chassis, body trim, etc.). Then came the technically more complex and capital-intensive power-train (engines, transmission, axles, and gear boxes), followed by heavy body stampings. Each of the major components, such as engines, went through successive stages of displacing imported components with locally manufactured engine parts. The easy, hang-on parts required modest investments in equipment and tooling; the more complex hang-ons, power train components, and body stampings entailed heavier investments in fabricating equipment and tooling. (A $220 million investment is needed for a typical complex to produce up to 200,000 units a year, for about 60 percent of the vehicle value.) This stage requires substantial commitments of managerial resources to phase in central plant production and suppliers industries. Sustained enterprise-to-enterprise relationships of three to eight years were required to implant production techniques among vehicle manufacturers and their supplier network. Logistically, complex systems to procure, package, and ship thousands of parts items to the dozens of automotive complexes located throughout the world were developed by the leading automotive firms.

Profits in developing-country markets, although initially lucrative, have eroded over time. Profit levels have been a function of the size of the domestic market, the level of effective protection (the difference between tariffs on imported inputs and the tariff on fully assembled vehicles), the stage of industrial development of basic materials and parts supplier industries, the nature and extent of the automotive decree (local content regime and its time phase-in requirements), and the number of firms, plants, makes, and models allowed to proliferate under the industrial regulations governing the automotive industry.

U.S. automotive firms have been generally adverse to making investments in countries with varying combinations of the following factors: relatively small and low-growth markets, the absence of uninterrupted tariff protection, a local-content regime requiring rapid phase-in of part manufacture (15 to 20 percent per year), and rigorously enforced price controls. The cost-profit squeeze has been felt in recent years in Mexico for the following combination of reasons: price controls prevail, its market is 60 percent less than Brazil's, the pressure on model changes is intense (proximity to U.S. market creates a demonstration effect), and domestic manufacture of local content (or equivalent value in exports) is now beyond the 75 percent mark. The prospect of future market expansion is important—in the early sixties, many investments (in Mexico particularly) were made with a view toward serving the Latin America Free Trade Association (LAFTA) market, but these expectations for regional markets largely failed to materialize.

More recently, certain developing countries have asked foreign firms to increase manufacture of automotive products for export as part of national efforts to rationalize production, by replacing high-cost local content with internationally competitive volume production. Ford's new engine plant in Brazil, which has a production capacity of 100,000 units a year, represents over $200 million in new investment. Developing countries have also sought, where feasible, to expand national ownership and control of industry, to more effectively screen foreign licenses and investments, and to expand domestic design and engineering capabilities.

Automotive Case Studies

The earlier section on automotive technology transfer requirements suggests the high sophistication and complexity of technology in this industry. The automotive cases we selected to study all involve the release of front-end technology: GM's proposed sale to Poland of design and production know-how for a new line of commercial trucks; Cummins Engine's sharing of the production function with the Japanese for advance diesel engines; Gamma Auto's sale of production technology to a socialist country for latest generation automotive parts; and Bendix's license to a German firm for the manufacture of an electronic fuel injection system.

Various considerations motivated the four firms to enter into these technology transfer agreements. General Motors and Gamma Auto saw an opportunity to earn corporate returns in markets that would otherwise be inaccessible to them. In addition, GM, due to the buy-back component of the contract, could have supplied a larger Western European market without capitalization in its own production facilities. Gamma Auto, recognizing that it can exert virtually no control over their client's distribution of the technology, has plans to commercialize the next generation of the manufacturing technology before the client is prepared to compete in international markets. The Cummins Engine Company was prompted to assign major manufacturing responsibilities to its former Japanese licensee due to severe capital constraints. Bendix was motivated to link up with a strong European partner in order to earn a return, to avoid patent infringement claims, and to benefit from technology exchange.

In discussing the purchasers' motivations, we can again group together the two cases involving socialist economies. In the GM and Gamma Auto cases, the purchasers wanted internationally competitive technology and production facilities. In addition the purchasers were eager for the training of technicians and managers in production design operations, as well as engineering, in order to develop indigenous operative and duplicative capabilities. In the other two cases, the purchaser was already quite advanced in technology and production know-how. Komatsu, confident that it could rapidly absorb the technology, was anxious to assume responsibility for a major manufacturing role to enlarge its world markets. Bosch, Bendix' licensee, sought to maintain its technological parity with the U.S. firm through the cross-licensing arrangement. Furthermore, the license enhanced Bosch's ability to serve European firms whose automotive exports to the U.S. were required to meet certain environmental standards.

All four agreements contain implications for the U.S. economy. GM and Gamma Auto have directly or indirectly contributed to establishing a potential competitor in industrially advanced country markets. Output from these facilities may eventually displace U.S. exports of original equipment, automotive replacement parts, or complete vehicles. This development will have long-term adverse consequences for U.S. jobs and income. The Cummins-Komatsu agreement results in an immediate erosion of U.S. production jobs yielded to Japanese production facilities. It also represents a marked shift in emphasis by the firm from production and investment toward marketing and research. The Bendix-Bosch interchange is advantageous to the relative market positions of involved U.S. and German firms, but it also enhances the competitive position of foreign manufacturers vis-à-vis other U.S. firms.

Case Study: General Motors' Proposed Truck Manufacturing Agreement with Poland[a]

A general lackluster record of earnings on its foreign investment base coupled with setbacks in the domestic market due to the energy crisis and the

---

[a]At the time of writing, the agreement was still under negotiation, pending the securing of international financing. All efforts subsequently failed, including a loan request to the U.S. Export-Import Bank, and General Motors withdrew from the negotiations. The information contained in this case study was obtained from sources other than GM because the corporation was prohibited from disclosing any information about the contract negotiations.

economic recession has led the General Motors Corporation to reevaluate its global policies. GM no longer considers equity and managerial control indispensable to its overseas operations, and it is now willing to release technology and production know-how to noncontrolled affiliates in order to earn a return on these corporate assets. Its negotiations with the Polish government to provide manufacturing know-how for a new line of trucks serves as an excellent illustration of GM's new corporate philosophy on how best to earn a return on corporate assets. Even more significant, under the proposed arrangement in Poland, a noncontrolled affiliate would have become part of GM's international supply system—which again represents an unprecedented step. It reflects, in part, capital shortages to finance global growth and development (GM borrowed $600 million for use in its worldwide operations in 1975) and the aversion toward risk and uncertainty in new capital investments in many parts of the developing world.

The negotiations in Poland also represent one more step toward the internationalizaton of the production system and the product design cycle, beginning with GM's early investments in Great Britain, Germany, and Australia. The two basic reasons for decentralizing these functions were to accommodate more effectively local market demands and production requirements and, more recently, to reduce production costs for major components in response to intensified competition at home and abroad.

Under the proposed production and marketing agreement with the Polish government, GM was to design a series of vehicles, ranging from light vans equipped with four-cylinder diesel engines to 5-ton trucks. The trucks were to be production-engineered for a small truck facility, the Fabryka Samochodow Ciezarowych in Lublin. The Lublin factory, which currently produces Zuk (Beetle) small trucks of 1800-pound capacity, equipped with the obsolete 2000-cc engine of the Warszawa automobile, was to be expanded and modernized according to GM specifications and design.

The broad terms of the plan called for Poland to purchase required equipment from Western Europe. The unique aspect of the proposed GM-Polish agreement was that the vehicles to be manufactured at the plant were to be of an original design meeting Polish specifications and not an existing GM design. GM was developing the production documentation at its Vauxhall facility in Britain with the participation of Polish technicians. Approximately 100 GM design engineers from various parts of the world corporation were to be assigned to work with and instruct their Polish counterparts.

The Lublin group was to obtain from GM complete documentation on production methods, including physical dimensions of components and parts, materials specifications, process sheets detailing fabrication steps, tooling and machining requirements, and quality-control and testing procedures. The Polish officials anticipated reaching an annual production volume of 100,000 units and exporting a percentage of the new units to hard currency areas. Export earnings would help pay for GM technology and component imports (from Great Britain and elsewhere) during the extended run-in period.

The proposed agreement signaled that the largest manufacturing company in the world would be expanding its base into Eastern Europe, since it expected to buy back between 15,000 and 18,000 of these vehicles annually. The vehicles, light vans, would have been sold in Western Europe. In turn, Poland would have been granted exclusive marketing rights within the member nations in the Council for Mutual Economic Assistance (Bulgaria, Czechoslovakia, East Germany, Hungary, Poland, Rumania, Bulgaria, and the Soviet Union).

The initial expansion was expected to cost about $600 million, and the total investment could eventually have reached as much as $1 billion. The U.S. component was estimated at $300 million initially. At the time, it was hoped that a substantial portion of the financing would come from the U.S. Export-Import Bank.

From the Polish standpoint, there were considerable advantages in coproduction and comarketing agreements, which they have been negotiating over the past few years with U.S. firms. A typical technical assistance agreement would entitle the Polish enterprise to advice and guidance on specific designs, applications, operations, and manufacturing problems. In addition, the firm's specialists would have made periodic visits to the local facility and trained technical personnel. Furthermore, GM would agree to keep its flow of technology up-to-date throughout the life of the agreement by furnishing information about engineering modifications and improvements in a particular product that it had actually put into commercial use. Thus, the Polish enterprise would have had not only an extended technical assistance contract but also rights to a Western trademark and a product purchase commitment. In providing rights to the trademark, GM, in effect, would have guaranteed that the Polish product would be as good in every respect as any product coming off its own production lines. The Polish enterprise also would have been assured the transfer and constant updating of those production methods and product design modifications that comprise the firm's latest technology. Product purchase commitments would have made the technology transfer a self-financing proposition, based on hard-currency export earnings.

The proposed agreement in Poland represented a radical departure from GM's traditional mode of involvement in overseas ventures. In the past, when transferring unique or proprietary technology abroad, GM took considerable safeguards to ensure that the know-how stayed within the domains of controlled affiliates and, preferably, in the hands—and minds—of its own engineers and technicians assigned to the overseas affiliate. What is striking, therefore, about the proposed GM-Polish agreement is that it represented a complete reversal of its policies concerning technology transfer. In effect, GM would have been leapfrogging the entire product life cycle by designing and engineering a prototype truck for the Poles.

The proposed agreement also reflects a significant change in GM's attitudes toward control of management and production facilities. This is another area over which the firm traditionally felt it must have unconditional command in order to maintain quality standards and, more important, its share of the world market. The proposed contract with Poland provided for coproduction and comarketing, that is, sharing—with a noncontrolled affiliate—responsibility for production, resources, manpower, returns, and corporate costs.

Finally, a third rather novel aspect of this commercial transaction is that it would have required virtually no capital outlay from GM—again, a radical change in its mode of foreign operations. The main cost to GM, approximately 100 technicians and managers on temporary loan, indeed, would have been a modest investment in terms of the anticipated return. A more significant element of this noncapital transaction with Poland, however, was that the new vehicles would have complemented and helped to fill out the GM line in world markets.

A longstanding concern of the Western European automotive industry is that the Soviet Union and Eastern Europe will one day flood their markets with cheap cars and trucks. While this threat has not yet materialized, Eastern European economies have changed direction radically in the last five years on the production of consumer goods, and their rapid move into the motorized age has been accompanied by a keen eye to exports. Eurofinance, a Paris-based research company, has estimated that East-bloc production, including that of the Soviet Union, will climb from the current total of 1 million trucks per year to nearly 1.5 million by 1980—the number of trucks Western Europe currently produces annually. They suggest that COMECON's motor vehicle industry will assign the highest priority to truck manufacturing and related equipment and parts through the 1980s. One sign of this new emphasis is the fact that Eastern European countries are building roads faster than they are building railroads.

Poland, a nation of 33 million people, has been a leader in this field since the ouster of Wladyslaw Gomulka five years ago. His successor, Edward Gierek, has pushed the country's economy deliberately in the direction of increased industrial productivity and the manufacture of consumer goods, including passenger cars and commercial trucks. Some 25 percent of all Polish engineering investment over the last five years has gone into the motor sector.

The results, in purely statistical terms, are impressive. In 1970, Poland produced about 60,000 cars. In 1975, the figure was a little under 200,000, and if all goes according to plan, it virtually will double again by 1980. Overall, then, Poland would have an automotive industry of at least a medium rank in comparison to the rest of Europe. Even in the West, such an intensive effort toward mobilization has been rare: in the planned economies of Eastern Europe, only the Soviet Union has made a comparable effort.

Foreign-exchange earnings (from truck exports) are important to the Polish economy. Poland now has a running trade deficit of $3 billion to $4 billion, the result, in part, of its efforts to modernize existing plants and develop new industrial capabilities. Poland also had an immediate need for trucks in connection with industrial growth and development. In September 1976, Steyr-Daimler-Puch signed a $250 million agreement with Poland to help build a heavy-duty truck engine plant, but in the meantime it will be selling heavy-duty, 16-ton trucks to the Poles. The Poles are also assembling Volvo heavy trucks and producing their own Star trucks (4-8 tons) at the Starachowice factory in central Poland.

Poland currently offers an extremely attractive environment for foreign commercial interests. It has developed a sufficiently sophisticated technological capacity to produce high-quality products. Moreover, it is reportedly able to do so at less cost than would be possible in the U.S. or Western Europe, primarily because of its still comparatively low wage scale. The Polish government is strong and quite stable and has a history of respecting business agreements. The country also has a well-developed infrastructure for industrial activity: abundant supplies of energy, good ports, and railroad systems.

The proposed contract represented more than $50 million in potential earnings to GM, aside from the sale of vehicle components and replacement parts to have been shipped to Lublin during the phase-in period. Other advantages that would have accrued to GM in the Polish agreement include: savings in capital investments of up to $200 million for the new line, sharing in design and engineering costs for the new model, and royalty earnings on all output, including sales to otherwise largely inaccessible Eastern European markets.

Implicit in the arrangement was the GM perception of the Polish enterprise as a partner who would be dependent on the company for an appreciable number of years. GM's purpose in establishing a dependent relationship with Poland was to ensure that a continued lag existed in Poland's design, production-engineering and marketing capabilities. A continuing dependent relationship between GM and Poland would have depended on the latter's technical and managerial absorptive capabilities, the speed with which they developed their own markets, and the length of time they felt they must rely on the GM trademark as a symbol of quality.

The new agreement in Poland would have had little impact on U.S. income and employment, at least for the next few years. During the phase-in period, most components and parts probably would have been supplied from GM in England and Germany or their local parts suppliers. Very few Polish imports of trucks have come from the U.S. in recent years, so there would have been little or no displacement in this regard. GM's earnings for technical services and royalties would have eventually added to U.S. foreign-exchange balance.

It is highly probable that the expanded truck plant would have begun exporting trucks mainly to the European market within five years. These exports would have competed with existing European and British auto manufacturers rather than displacing U.S. exports in these areas and, at least for the time being, would not have posed a foreseeable threat of entry into the U.S. market.

The technology that GM proposed to provide, although considered first choice by the Polish authorities, could alternatively have been supplied from European or Japanese sources. The Polish agreement would have provided GM with an opportunity to expand its international production and marketing in areas where it was otherwise excluded (Poland, other Eastern European countries, the USSR, and other, nonsocialist trading partners). It also would have afforded GM an opportunity for additional earnings in the world market from the sale of Polish trucks through GM marketing channels where it would otherwise not consider it worthwhile to invest in engineering new models and in additional plant facilities. Some additional earnings would also have accrued to GM's technology transfer agents (most of whom would have been recruited from non-U.S. sources). Most of the parts sales during the plant phase-in period would probably have been made from non-U.S. sources. On balance, there would have been a small gain to the U.S. economy and virtually no identifiable loss.

It is significant to note, however, that GM was willing to put a potential competitor into business, presumably on the assumption that the Polish enterprise would have operated in a market segment that GM could not otherwise reach through its owned and controlled production facilities. There also might have been an eventual displacement effect on vehicle and parts produced in the U.S. if GM had found it less costly and more advantageous to procure vehicles and components from Polish sources for its global system.

Case Study: The Cummins Engine Company's Agreement with Komatsu, Ltd., for the Manufacture of an Advanced-design Diesel Engine

In 1973 the Cummins Engine Company, the world's largest producer of diesel engines, negotiated an agreement with its Japanese licensing partner,

Komatsu, Ltd., for the production of a major portion of the components for its new K engine. The K engine was Cummins' newest-generation diesel engine—representing seven years of research—and was born in response to Cummins' perception of future demand in the power market in the 1970s. Specifically, it was anticipated that sizable markets would open up in the 400-600 hp and 800-1200 hp range, and the K engine has been clearly the most competitive product developed to date to meet that demand. It was largely Cummins' assessment of its strengths in the engine field, coupled with the necessity to conserve capital, that influenced its decision to source production of the most capital-intensive components of the K engine.

In the 1950s, the corporate goal of Cummins' management was a 15 percent annual growth in profit, which implied a doubling every five years and a 15 percent return on shareholders' equity. That Cummins realized these objectives in the 1950s and 1960s is evidence of its success during that period. In the late 1960s, however, projections into the future of the potential growth of the power market, and diesel power, in particular, indicated that demand would be less than in previous decades. Cummins' management estimated that no more than 8 to 10 percent of growth could be anticipated to come from this market and concluded, therefore, that the firm could not remain committed exclusively to the diesel engine business and continue to achieve these objectives.

Cummins' financial situation in 1969 was an equally compelling reason for making some alterations in its strategy and operations. Prior to 1965, it had been the company's policy to raise capital largely from internal sources; after 1965, this policy turned increasingly toward stock issues. With the substantial downfall of the stock market in 1968, the subsequent rise in interest rates, and a decrease in the amount of internally generated funds, Cummins began to seriously scrutinize its financial management with a view toward conservation of capital.

Cummins' corporate policy of capital conservation was reinforced by its assessment of the economic environment. At the time, Cummins' operations consisted of combining labor, capital, and technology to produce engines. A forecast of the behavior of these factors in the future indicated a rising cost of money and labor and an accelerated pace in technological innovation. The rising cost of money dictated that it be used only for high-return investments; increasingly expensive labor suggested its use only at the highest-potential-return level; and the speed with which technology was changing required maintenance of maximum flexibility at all times to adjust to continuous change. Combined, these three elements pointed to the conclusion that Cummins should leave the capital-intensive process of manufacturing to someone else and conserve its resources for high-yield operations, while maintaining its flexibility to adopt new manufacturing technologies.

It was generally agreed that the strength and prominent position Cummins enjoyed in the diesel engine market was derived largely from its technological leadership in the engine field and the efficient network of distributors it possessed to service the ultimate customer. The actual manufacturing of the engine—however demanding in sophisticated skills—could be done equally well by other producers. Assigning production to another manufacturer would free a substantial portion of Cummins' resources for concentration in those areas in which it enjoyed an advantage over other producers in the market—designing and marketing engines.

A basic assumption throughout the seven-year period of designing and engineering the K engine was that its production would be located in the U.S. In 1970 gross estimations of the cost of building a new plant in the U.S. to produce the engines and of the market the K engine could be expected to command indicated a rate of return considerably higher than the company's required fifteen percent—approximately 32.1 percent.

Confidence in the new K engine project was severely shaken, however, when it became apparent that an investment of $44 million would be required from 1972 through 1985 in order to produce the engine in the U.S. This figure was based on capital requirements for K engine production that Cummins' present plants and equipment could not meet. Calculations of material and labor cost per unit took into account present costs estimated by the research department and assumed a learning process in which costs per unit of output would decline over time. The cumulative capital investment for the period came down to $39 million when the calculations included the capital saved due to a decrease in maintenance capital for its ongoing manufacturing operations for the NH and V engines.

In late 1971, a more detailed analysis of the rate of return Cummins could reasonably anticipate by producing the engine in the U.S. indicated a slightly lower rate of return—26.4 percent with a payback period of 7.8 years (by 1979). The figure took into account the rate of inflation in the U.S. and new regulations on investment tax credits and accelerated depreciation. The most sensitive assumption in the analysis was the selling price. A 25 percent decrease in the selling price would bring the rate of return in the project down to 3.9 percent. Almost equally sensitive was an expected increase in costs of 25 percent, which would result in a return on the project of only 14.3 percent.

The size of the requisite capital outlay coupled with the uncertainties in returns attending production of the K engine in the U.S. prompted Cummins' management to revise their plans and consider alternative production locations. In their deliberations, Cummins' management focused on two seemingly unrelated factors. First, Cummins was an international company committed to worldwide participation, and second, its largest competitor in the off-highway market, Caterpillar, was beginning to make inroads into the U.S. automotive market. Combined, these two factors directed attention to one of Caterpillar's largest international competitors and one of Cummins' long-standing and major licensees, Komatsu, Ltd., in Japan.

Through ten years of license agreements to manufacture the NH engine, Cummins had developed considerable appreciation for Komatsu's ability to undertake sizable orders for engine manufacturing and to produce engines that met Cummins' high standards. Komatsu is a vertically integrated equipment manufacturer with production facilities at least as modern as Cummins' facilities. Komatsu's foundry, in fact, is far superior to those commonly found in the U.S., and its machinery operations employ technology as advanced as that used by Cummins. Labor-capital ratios are quite comparable to those at Cummins. Finally, Komatsu's corporate management was highly respected by Cummins for its sharp perception of market forces, its success in differentiating its products from top-level foreign brands through high quality, and its boldness in entering new markets, most notably the Eastern European market. They also had succeeded in penetrating the Chinese (PRC) market.

The terms of the contract called for the production by Komatsu of eight major components of the six- and twelve-cylinder versions of K series engines. Those eight components were selected on the basis of their high capital intensity. If they were produced in the U.S., they would require 77 percent of the total capital investment through 1976. Additionally, they contain 44 percent of the total material content and utilize 31 percent of the total labor content of the base K engine assembly. At present, Cummins purchases the eight components from Komatsu and carries out final assembly and testing of the engines in its Charleston, South Carolina, and Daventry, England, facilities. In all likelihood, the complete engines will eventually be manufactured by Komatsu and purchased by Cummins at a prearranged price.

Cummins' decision to share the production function of its new generation of diesel engines with Komatsu arose largely out of its severe capital constraints—a not uncommon malady suffered by many U.S. firms in recent years. In fact, to speak of the agreement as an alternative to completely manufacturing the new engine in U.S. facilities would be euphemistic; in all likelihood, Cummins could not afford to produce the K engine without such a strong—financially and technically—partner as Komatsu. Therefore, a major benefit of the agreement for Cummins is the ability and means with which to introduce into the market a highly competitive product for which, according to all indicators, there will be a large demand. In addition, by relieving itself of producing all but 20 percent of the K engine, Cummins took a major step toward implementing its new strategy of focusing its resources on technology and marketing rather than manufacturing.

The agreement afforded Cummins the opportunity to strengthen its relationship with Komatsu. This relationship in the past had provided Cummins entry into the Japanese market, and it was expected that it would open other valuable world markets in the future. The relationship had also earned for Cummins royalty payments on engine manufacturing in Japan. Under its earlier licensing agreement, Komatsu paid Cummins a 5 percent royalty on the manufacturing cost of sales of all licensed products manufactured or assembled by Komatsu and used to power its own equipment. In addition, the two firms established a Komatsu-Cummins Sales Company, Ltd., in the early 1960s, of which Cummins owns 51 percent. All sales of parts and engines to other Japanese manufacturers, as well as sales of parts for Komatsu products, are handled through the company, which also serves as Cummins' exclusive sales agent in Japan for parts and engines from Columbus and other Cummins' plants. In all likelihood, the sale of parts and components of the K engine as well as of the complete engine in later years will also be handled through the Komatsu-Cummins Sales Company.

The major benefit enjoyed by Komatsu, Ltd., as a result of the agreement was further infusion of very sophisticated technology. In its earlier licensing arrangement with Cummins, Komatsu acquired a vast array of technical knowledge that allowed it to achieve considerable success in some of its machinery. Komatsu was manufacturing a quality-equivalent Cummins diesel within less than two years after the license agreement had been signed.

In the manufacturing of a diesel engine for commercial trucks, there are approximately 750 parts, ranging from cylinder blocks to fuel injector pins. Each part may require anywhere from 5 to 75 separate process steps to produce a finished part. In the U.S., close to 200 plants supply Cummins with materials, raw castings, forgings, components, and parts. To produce these parts, as many as 300 different materials are required, and there are narrow standards on physical and chemical characteristics and shapes of finishes for each of them. Approximately 15,000 manufacturing steps are required to convert materials and castings into finished parts for a single engine model. To be involved from the very beginning, and in such an extensive manner, in the manufacture of the newest generation of diesel engines is indeed a great boon to Komatsu's already sizable reservoir of technical and production-engineering expertise. The agreement can also be expected to expand Komatsu's already established link to the U.S. market.

The U.S. economy does not fare quite as well from the agreement as do the two signatories. It is true that capital constraints eliminated Cummins as the major manufacturer of its new engine and that Komatsu, given its former relationship with Cummins and reputation in the field, represented a logical candidate. Nevertheless, the fact that the bulk of the production is sourced from Japan translates into a loss of jobs and income for the U.S. Equally important, it means that the know-how for the newest generation of a highly sophisticated U.S. product has been released to a foreign, non-controlled affiliate. And, in this case, the affiliate has already demonstrated itself to be fully capable of absorbing advanced technology as well as astute and successful in entering new markets.

It is not at all clear whether Cummins would have entered into the agreement had it not suffered from such severe capital shortages or had there been a suitable merger arrangement that would have furnished the additional capital required. The problem created for American firms by the U.S. capital market and antitrust laws plays an important role in such decisions and requires close examination. Unfortunately, these topics are not within the scope of this paper. It is clear, however, that the agreement has resulted in a further erosion of the U.S. production and technological base.

Case Study: Gamma Auto's Agreement with a Socialist Country for an Auto Parts Manufacturing Facility[b]

In March 1975, Gamma Auto completed negotiations with the state trading corporation of a socialist country for a sizable contract under which the firm will supply manufacturing equipment and technical know-how for an automotive parts manufacturing facility. The contract calls for the training of the client's technicians and engineers by Gamma Auto in their U.S. facilities. In addition to receiving instruction in the operation, engineering, and maintenance of the plant equipment (with complete access to all manuals, drawings, specifications, blueprints, process sheets, etc.) and management, the trainees will also be taught to be instructors when they return home. The equipment Gamma Auto will procure and sell to the client for the new facility—including metallurgical, mechanical, electroplating, and two casting lines—will be American-made and the most advanced available.

Former licensees of Gamma Auto—one British and one German—as well as a couple of U.S. firms were Gamma's competitors for the job. The precise technology contracted for is by far the most sophisticated available and incorporates the highest-premium material. Older technologies might have served their purposes equally well, but their client wanted the "Cadillac line." This was the case for a variety of reasons. For a socialist economy, the market nuances are of marginal importance; the determining factors are that the product is functional, that its production is efficient and lends itself to high volume, and that the production operation has a long life expectancy and can be smoothly managed. Equally important is that the experience of operating an efficient, functional, and high-volume manufac-

[b]Gamma Auto Corp. is a pseudonym. Anonymity was granted this firm in exchange for sensitive information on the case.

turing facility may, in time, impart greatly needed design and engineering capabilities to the operators. That Gamma Auto was the first to develop the materials technology, remained in the forefront of technical development, and enjoyed an excellent reputation internationally suggested to their client that they could provide them with a facility whose output and operation best met these criteria. Gamma's client and other socialist countries have also demonstrated in recent years an increased appreciation for resource conservation and capital-cost efficiency, both of which would be best achieved by acquiring the technology from Gamma.

Such explanations for the client's behavior in purchasing technology assume a rational model of decision making. Insistence on the newest, most sophisticated technology available, however, may represent a self-conscious reaction against the antiquated machinery and processes they have employed for years and a conscious effort to appear to be riding the wave of the future. It may also be a symptom of this country's bureaucratic aversion to risk taking. If the project is not successful, the state official at least cannot be held responsible for selecting second-best equipment or technology. An indication of the negotiators' anxiousness to obtain the best technology available was their final and last-minute acquiescence to Gamma's $6 million increase in the price of the contract—ostensibly to cover anticipated major increases in infrastructural costs.

Gamma Auto's long four years of negotiations with this country were instructive, if frustrating. It was their first commercial venture with this particular country, and they found their client to be an astute and skillful negotiator. Participating in the planning and negotiating sessions were so-called "pricing experts," that is, economists who had acquired considerable understanding of the U.S. economy during assignments here. They had a detailed familiarity with current relevant U.S. labor statistics, GNP, inflation rates, cost of living, and so on, and called upon Gamma negotiators at several intervals to justify the costs of the component parts in the contract according to such indicators, in an effort to reduce its total cost. On several other occasions, at points when Gamma negotiators felt that talks were progressing well, their client would press for a reduction in the total cost of the contract. They attempted to make technical changes in the contract during the final negotiating stages. Gamma had limited information on the client's negotiating frame and other related conditions prior to entering into the negotiations, and their inquiries to U.S. government officials, trade associations, and banking institutions yielded little useful or reliable information.[2]

This particular socialist country is well aware that it is perceived in such circumstances as a powerful state trading company and is, therefore, sensitive to accusations that it can exploit its position as a monopoly by playing off the private companies bidding for contracts. In addition, it represents a potentially strong competitor in the same world market once its acquisitions come on-stream.

Despite the disadvantages Gamma Auto suffered in the negotiations due to lack of information and experience, it fared quite well by the final terms of the contract—as evidenced by the fact that its original bid in 1971 was for $15 million and the final price agreed to was a multiple of this figure. In addition to making a down payment, the client is obligated to pay the full amount at the time the goods are delivered to the common carrier, minus a guarantee fee of 5 percent.

The ostensible reason for more than a 200 percent increase in the final sum of the contract was that its terms called for slightly more equipment and technology than were under negotiation in the initial phases of the talks. Gamma officials, however, indicated that the real reason for the price hike was that they were monetizing their fear of being unable to accurately anticipate what the total cost of the contract would be. It is the policy of the purchasing country to prohibit the inclusion in contracts of escalation clauses to cover inflation.

In addition, Gamma feared that their client might share the new technology with other socialist economies. While ensuring that the client could not sell the automotive parts in their own markets by inserting a patent rights clause in the contract, Gamma could not restrain them from sharing the technology if they so desired. Earlier Gamma had seen their own patented chemical process used in (chrome) plating equipment copied in another socialist country from a West German facility they had licensed. This experience and others led them to the conclusion that it would be futile to attempt to legally restrict their client from sharing the technology with other enterprises. Gamma was also concerned over the contribution that an efficient, sophisticated auto parts manufacturing facility might make to the client's military capabilities as the parts could be used not only in trucks or autos but also in military equipment.

After the first year of negotiations, the firm informed the U.S. Departments of Commerce, State, Treasury, and Defense and the White House of the status and substance of the talks in order to find out whether they had any objections to the sale of the technology. (It did not occur to them to contact the Department of Labor.) None of the government officials consulted expressed objections or reservations on either economic or national security grounds. The existence of two alternative and willing foreign sources of essentially the same technology seemed to be a major consideration in the government's assessment of the proposed transfer.

In interviews, Gamma Auto officials expressed concern over the institutional inertia they encountered in Washington. Of the firms we interviewed, Gamma was the first to articulate such a sentiment. "What this country needs is a Japan, Inc. model," one Gamma official suggested. The Japan, ᵢnc. model refers to the unique relationship that has developed between the Japanese government and that country's private sector based on mutual cooperation and benefit. The government issues administrative guidelines to private enterprises and its strongest means of restraint is the withholding of subsidies to foreign investment. Extensive consultation takes place on all foreign business transactions, and both parties systematically consider the commercial, financial, political, and labor ramifications of each particular transaction. Japanese multinational officials have a deep respect for the symbiotic relationship that exists between a healthy, dynamic domestic economy and a healthy, dynamic international firm.

Gamma, in addition to negotiating an extremely lucrative contract, has further insured itself against incursions into their international markets by withholding from their client information on their newest product technology, which might be introduced to the market in the future. Based on a different technology, the material costs of the new product could be significantly less than that of the auto part to be produced at the client's facility, and the finished product will be much cheaper than the price of the client's product. There is an implicit assumption here that the client's absorptive capacity will be extremely low and that considerable time will elapse before it can enter export markets. To date, this is not an unrealistic estimate, in light of the year delay that already exists in the client's completion of the plant.

Further evidence of anticipated delay is the arrangement Gamma insisted on for a warranty of the project. They avoid referring to the agreement as a turnkey operation and refused to provide a guaranty to their client on the performance of the manufacturing facility because they feared they would be committed to supervision, maintenance, and operation of the facility for years to come. Instead, Gamma has guaranteed the equipment, technology and the manufacturing operation for eighteen to twenty-one months from the day of the equipment's delivery or for twelve months after start-up of the facility, whichever comes first. Gamma also anticipates that it may be necessary for the client to come back to them for additional tools, equipment, and training not called for in the original contract.

Gamma Auto officials have succeeded in negotiating a contract that places them in a clearly advantageous position vis-à-vis the client—if events occur as they anticipate. Having sold the technology needed to produce highly competitive auto parts and recognizing that they can exert limited control over the client's future dissemination of the technology, Gamma is compelled to complete development of the newer auto part and have its production on-stream and successfully commercialized before the client has reached a comparable stage with the current technology. If Gamma has underestimated the amount of time required to commercialize the new auto part, it may lose some of its market share, particularly in Western Europe, to their client.

The likelihood of this scenario actually materializing, however, is slim, given Gamma's present attitude toward technology. It management refers to Gamma as a technology-oriented company and is committed to the idea that continuity in an R and D effort is essential to its long-term growth. Its budget for R and D is kept at a sustainable level and does not fluctuate in response to year-to-year economic conditions or earnings pressure. Gamma's internal spending for research and development has increased from about $10 million in FY 1971 to over $23 million in FY 1975. An additional $7.4 million for R and D was funded in fiscal 1975 by various government agencies. Total R and D in 1975 was $30.5 million on sales totaling $772.9 million, compared to $27.7 million in 1974—an increase of 10 percent. Its 1976 budget calls for total R and D of $35.5 million, or an increase of 16 percent.

Two concepts inform Gamma's approach to product development: the "integrated technology" concept and the concept of "market pull and technology push." The former concept means that the basis for the company's technical growth is formed by the integration of its closely related technologies—electrochemistry, electromechanics, electronics, and metallurgy. The technology of any one Gamma business is scrutinized for its potential to profit another segment by interchange of technical knowledge. The concept of "market pull and technology push" refers to the pull from customer problems and marketplace needs and the push from new possibilities for products arising from advancing technology.

The company has recently set up a New Business Division, which acts as an incubator for new products that may spawn new market areas for the company. This division develops products not within an existing division's markets, establishes a business plan, and actually begins a business operation that will eventually become a new corporate division. Projects for the New Business Division must meet these rigid criteria: a promise of market introduction within three to five years; potential total minimum market at $50 million, growing at 15 percent a year; pretax return of 30 percent on sales and 40 percent on investment; and potential to establish Gamma as a technical or market leader.

This clearly defined and detailed policy on technology and new product development (and understanding of the relationship between the two) distinguishes Gamma Auto from most other U.S. firms. In large part, it is this clearly defined policy that allows Gamma to be confident that its new auto part will be commercialized before its clients are prepared to enter export markets with the current technology. Should Gamma find that for some unforeseeable reason it must divest itself of particular auto parts manufacturing altogether, however, the U.S. industry itself could suffer increased competition.

A more likely scenario is that Gamma Auto, as well as other auto parts manufacturing firms, will face increased competition in developing countries, in markets that do not necessarily require the most advanced generation product. Their client has already approached the Indian auto parts market using technology acquired from the U.S. Furthermore, Gamma's sale of the production technology will preempt it and other U.S. manufacturers from making future sales of the particular parts, as end products, to the socialist countries of Eastern Europe and the Soviet Union, once the new facility is on-stream.

Case Study: Bendix' Licensing of Electronic Fuel Injection Technology to Bosch

The Bendix Corporation is one of a handful of first-tier suppliers to automotive original equipment markets (OEMs). Sales of automotive equipment for inclusion in original equipment are made to a relatively small number of customers, which is not surprising, given the high degree of concentration in automobile production. Total worldwide sales for original equipment and after-market use to American Motors Corporation, Chrysler Corporation, Ford Motor Company, General Motors Corporation, Peugeot, S.A. and Regie Nationale des Usines Renault accounted for approximately forty-one percent of the net sales of the firm's Automotive Group for the year ended 30 September 1974. (It should be noted that sales to GM and Ford in particular are made on a worldwide basis, following the multinational pattern of these OEMs.) Partly as a result of automotive OEM patterns, Bendix has shifted from being a domestic company with an international outlook to a true multinational company, in the sense that it coordinates its worldwide resources to tap additional world markets.

Historically, the company developed its licensing activities more or less as an afterthought, regarding licensing as an extra profit opportunity over and above the original domestic market for which a product was developed. However, as the automotive OEMs have become increasingly multinational in their mode of operation, the automotive market has changed dramatically, forcing first-tier suppliers to model their own operations according to those of the OEMs and respond to them worldwide. Increasingly, therefore, Bendix has adopted the policy not only of widespread licensing but of direct investment on a worldwide basis. In its licensing, the firm has a strong preference for an owned and controlled affiliate, but where the foreign party is particularly strong in a technological or market position, Bendix is agreeable to licensing unaccompanied by equity or managerial control. This was the case with Bosch, and the relationship that developed between the two firms will be the focus of this case.

Licensing of EFI Technology to Bosch

European companies have characteristically been much quicker to put new technology into production than the United States, where the concentration of industry and scale of manufacture have created a great deal of inertia. Partly for this reason, even though Bendix had pioneered the development of electronic fuel injection (EFI) for automobiles, it was in Europe that a mass market first developed.

Bendix did a great deal of development work on EFI in the 1950s which led to the publication of papers in the *Society of Automotive Engineers Journal* in 1957. This was pioneering work in the whole concept of EFI to replace mechanical carburetion. By 1960, Bendix had filed a base patent and subsidiary patents for various elements of an EFI system. The main elements of an EFI system are the following:

> *fuel delivery*: pump, injectors, pressure regulators, fuel manifold;
>
> *air handling*: throttle body with mixer, air passages, hot engine idle flow, cold-start system;
>
> *electronics*: electronic control unit, sensors in manifold, temperature sensor, speed sensor or "trigger," throttle position switch

The effort of the late 1950s was conducted cooperatively by several divisions of Bendix. Although company officials have no accurate figures on the level of effort, it seems possible the scale was on the order of at least 60- to 70-man years between 1955 and 1960. Initially, in 1958 and 1959, Chrysler Corporation showed a great interest in the concept and ordered EFI systems for 300 prototypes which were operated in a controlled fleet until 1961. This system was primitive, employing vacuum tube rather than solid state technology. Because of cost and durability problems—especially evident in poor performance in rainy or damp weather—the system was shelved by Chrysler in 1962. The subsequent view of Bendix was that the system had been brought to market much too early.

Three of Bendix' divisions were active in the early development of EFI—Bendix Research Laboratories, Eclipse Machine Division, and the South Bend (automotive products) Division. In each of these divisions, several people, perhaps an average of four or five in each group, were involved on a full-time basis. At Bendix Research Laboratories, for example, there was an injector task force and the South Bend Division was in charge of energy control aspects. In addition to the three main divisions, electronic specialists from the navigation and control group at Eclipse Pioneer and avionics specialists from the Baltimore Divison were also involved. Consequently, the main characteristic of the late 1950s is one of intense commitment to this potential new business area.

By 1965, Robert Bosch G.m.b.H., one of Europe's principal vendors of automotive equipment, began working independently on an EFI system. Although it had applied for a patent about the same time as the original Bendix patent, they became increasingly concerned by the possibility of patent infringements since Bosch systems were introduced on cars exported to the United States from Germany. The effort of Bosch shifted to high priority in the second half of the 1960s when Volkswagen came to them with an urgent request to develop EFI for several of its models. The original interest of VW was an EFI system which would incorporate the physical limitations of the VW bus, with its confined engine compartment. Eventually, however, the Bosch EFI system came out in 1967 on the model 411 VW fastback and squareback, but in the beginning only with some models. By that time, the primary concerns of VW were fuel economy and better engine performance. Eventually VW's main concern with EFI in the United States would be emission control. But the market development of EFI in Europe began with the need for getting the maximum economy and performance out of the smallest possible engine, given the trend in several European countries to scale registration costs to the displacement of the engine.

From the late 1960s through late 1975, Bosch in Germany alone had built over 3 million EFI systems for VW, Saab, Volvo, Porsche, Audi, and Renault. Current production in Germany is probably about 40,000 units per month, at a unit price to the OEM of about $150. A reasonable estimation is that 10 to 15 percent of European-built cars are equipped with Bosch EFI. To date, only a few thousand units annually are being built by Bendix in the United States. (The unit is standard on the Cadillac Seville and optional on other Cadillac lines.)

The key to Bosch's early concerns were probably the Bendix publications in the late 1950s and the patent filing in 1960. Even though Bosch had done a great deal of development in sub-circuitry for EFI systems that was completely original, especially with regard to the cold-start and warm-up circuits, they apparently felt compelled to approach Bendix concerning a license. During 1967 and 1968 there was strong interplay between the two companies to establish a mutually agreeable position in this new technology, an area in which the two companies had an absolutely dominant technical position. A license was issued by Bendix in 1968, followed by a technology exchange agreement in 1969. The technology exchange agreement gave Bosch relatively free access to engineers in Bendix, a regular interface on an as-needed basis determined by the engineering staffs, and frequent interchange both in technical meetings and by Telex.

With some support of Bendix patents and engineering data, Bosch developed three successive systems—the D, the L, and the K-Jetronic systems. In essence, the succession of L and K systems had depended on improved methods of sensing mass air flow. The D system, a so-called speed-density system, based on the measurement of manifold air pressure, was the first developed, primarily for the early models of the VW squareback and fastback. This was more recently replaced with the L system, which provides continuous sensing of mass air flow rather than manifold pressure as in the D system. Currently, Bosch is introducing the K system on 1975 VW models, which returns to a partial mechanical sensing and computing system. Given the degree of interchange that has occurred between the two companies, oversimplifications are risky; it is fair to say that the D system was structured along the line of Bendix technology but that subsequent systems are based more on Bosch developments.

European and Japanese automotive exporters to the United States often tend to regard the Environmental Protection Agency's regulations as a form of a nontariff barrier and, consequently, view an affiliation with an American partner as one means of overcoming such a barrier. Obviously, a licensee such as Bosch can be a far more effective vendor to Volkswagen, Volvo, Saab, and others if it has a direct pipeline to Bendix on the subject of emissions standards. Such assistance can be regarded as an element in helping the European and Japanese OEMs to penetrate the U.S. market.

Bendix personnel noted that in addition to the difference in emphasis between emission standards and fuel economy, the end use of American EFI systems in large V-8 engines versus European use in four-cylinder engines has also, for the time being, created divergent approaches.

Some Bendix personnel believe that, in the last three to four years, EFI technology flow has changed direction and now runs, in some instances from Bosch to Bendix. In the past year, for example, about a dozen Bendix engineers have spent periods of time ranging from one week to one month at Bosch to study specific component design, especially injectors, and the Bosch L-Jetronic design. There has been a fairly open exchange based on the cross-licensing. The Bendix EFI group has sent numerous delegates to Bosch. In general, they have followed the Bosch approaches to component design in principle but not in design detail.

Bendix spokesmen were reluctant to discuss specific licensing terms, apparently due to the constantly changing relationship and a forthcoming patent expiration. In general, however, when asked about licensing terms, Bendix officials indicated that, if a broad cross-section of American industry were polled, one would find that the average goal is a return to the licensor equal to about one-third of the profit of a well-run, well-established licensee with a broad market. In Bendix' international licensing, where highly industrial, high-volume licensees are usually involved, the licensing staff usually negotiates a royalty fee ranging from 2 to 5 percent of gross sales for automotive product lines, 5 to 7.5 percent for aerospace lines, and up to 10 percent in very specialized situations, especially in the aerospace field. Bendix considers itself as extremely liberal in front-end load compared with other American companies, ensuring only that the various routine front-end costs involved in preliminary analysis, negotiation, and early assistance are recovered.

In some cases, they do, of course, press for some kind of additional premium payment, especially if it is suspected that the potential licensee does not have a broad market or is merely attempting to bottle-up a technology. They are also careful to keep the amount of consulting and technical support that is actually specified within the license itself to a bare minimum, so that they will be able to charge, on a time and expense basis, for specific consulting tasks in the course of the license. In the final analysis, Bendix officials believe that license income must eventually be calculated on a per piece basis. Consequently, their primary concern is to determine whether or not the licensee can actually perform what is intended by both parties and then negotiate a royalty fee based on volume production.

Although specific details were not made available, it seems reasonable to assume that Bendix has received royalties on the order of 2 to 3 percent of Bosch's gross sales of EFI systems since the late 1960s. Although this is merely a conjecture, it can probably serve as a reasonable basis for planning and analysis in the present study.

Turning now to the interests of Bendix in EFI development, it is clear that electronic fuel management and electronic engine control are, in the view of many industry observers, the next major technological growth areas in the industry, due to the simultaneous pressures of better fuel economy and strict, but still undetermined, emission standards. Several different technical approaches are being developed, of which the Bendix EFI system is only one. However, Bendix regards the market as sufficiently promising to justify the establishment of its Electronic Fuel Injection Division, even though the present market is still relatively small.

Given the growing promise of the U.S. market and the more extensive production of the licensees, Bendix appears to have turned increasingly to Bosch both as a source of technology and as a component supplier. In the current production program, Bendix is using many Bosch components including pressure regulators, pumps, injectors, and electronic control units. Furthermore, in developing their own systems, Bendix has repeatedly started with Bosch injector designs, which are very highly regarded.

As for future trends in the technology, there is a current move from analog to digital electronics in converting sensed data. This is an important evolutionary change in the electronic control unit, permitting additional control functions and the incorporation of LSI technology. It is already clear that Bosch and Bendix are headed in the same direction.

One Bendix planner for automotive electronics sees the Japanese as major future competitors, pointing out that Ford and Toshiba are already collaborating in automotive electronic development programs and that Toshiba is also working on micro-processors for fuel controls. In fact, Toshiba is already one of three or four favored suppliers to Ford for alternators, regulators, and rectifiers. A Bendix and Bosch licensee and affiliate,

Diesel Kikki and a spin-off (Tidoshi Kikki), has entered the market, and a Japanese unit is already incorporated in the Datsun 280Z. In addition, Hitachi is now moving toward the production of entire EFI systems.

As the market starts to expand, both Bendix and Bosch are developing worldwide marketing plans. The view of engineers, not speaking for corporate management, is that if relatively standard requirements emerge in the different countries, especially based on the trend to smaller cars in the United States, then a great deal of cooperation would result.

Based on this single case study, and without recourse to a broader set of data, some preliminary assumptions and conclusions about the transfer of technology from a highly qualified U.S. manufacturer of automotive components to overseas manufacturers might include the following:

1. Given the increasingly multinational nature of the automobile industry, especially the worldwide coordination of assets to expand world markets, there is strong preference for licensing to controlled affiliates with common corporate objectives. This preference also suggests that licensing fees per se, independent of equity return, may not be regarded in the automotive components industry as a sufficient means of return on technological assets.

2. Licensing is, at least at Bendix, regarded as a profit center, and the licensing staff is gaining a stronger voice than it had previously in corporate product development decisions.

3. Licensing to noncontrolled firms abroad is accepted, when necessary to reach a major market, when the policy of a local government precludes equity participation or when the licensee has technical and market strengths that supersede concerns over lack of control.

4. Control over the flow of technology to the licensee is increasingly difficult to orchestrate, given such factors as the growing commonality of automotive technology and design cycles and the effects of governmental or supranational agency pressures.

5. Direct sales of automotive components are rarely a viable alternative to licensing. Due to factors of scale, transportation costs and local pressures to retain manufacturing value added, direct sales are usually limited to aftermarket products. Supply of OEM components is normally provided through licensed manufacture and usually requires further product development by coordinated international elements of the company.

Based on this single case study, it could be assumed that the search for a relatively simple and uncomplicated set of license relationships, among U.S.-based multinational, high-technology firms, might prove to be both frustrating and misleading. It seems more likely that a realistic model of the technology transfer process must be one which accommodates a great deal of complexity in terms of technology flow patterns and compensation for the sharing of technology.

Notes

1. The statistical information contained in this section comes from Motor Vehicles Manufacturers Association, *World Motor Vehicle Data*, 1975; and Motor Vehicles Manufacturers Association, *1975 Automotive Facts and Figures*.

2. The exceptions were two individuals, one of whom worked at McKinsey and Co. and had negotiated over twenty contracts with socialist countries and the other served as commercial attaché at the U.S. embassy in the country.

4 The Computer Industry

Sector Overview

Significant advancement of computer technology did not occur until the middle of World War II. By the war's end, a small number of operations research analysts in Germany, Great Britain, and the U.S. utilized primitive computers to analyze wartime problems. Much of the early development of the U.S. computer industry was funded by Department of Defense research contracts.

In 1951, Remington Rand became the first U.S. corporation to sell civilian computers. Four years later it merged with Sperry Gyroscope Corporation to form Sperry Rand, a firm that dominated the computer market until 1956, when IBM surpassed it in commercial sales. IBM rapidly consolidated its position as a result of its combined research and development, production, marketing, and service capabilities. By the end of the decade, the U.S. industry was humorously described as "IBM and the seven dwarfs," the "dwarfs" being Burroughs, Control Data, General Electric, Honeywell, National Cash Register, RCA, and Sperry Rand. During the 1950s and 1960s, dozens of other U.S. firms attempted to gain a significant foothold in this industry. However, prior to the mid-sixties, few firms other than IBM earned profits from their computer operations.

By 1966, IBM had obtained more than a 70 percent share of the computer market in the U.S., West Germany, France, and Italy. At the same time, it had penetrated more than 30 percent of the Japanese market and over 50 percent of the British market. Although IBM's position in these markets has weakened somewhat in recent years, it is still the dominant firm in the world industry. In 1975, its sales volume exceeded $14.4 billion, making it the seventh largest corporation in the U.S. At that time, the annual sales volume of the world computer market was approximately $20 billion.

Due to IBM's competitive strength, several important U.S. manufacturers have chosen to withdraw from this business—General Electric in 1970, RCA in 1971, and Xerox in 1975. Other firms have opted to compete with IBM solely in the production of special-purpose computers or computer peripheral devices.[1]

In recent years, manufacturers in the U.S. and Western Europe have begun to emphasize peripheral devices that are compatible with IBM central processing units.[2] These firms attempt to provide IBM customers with lower prices or performance advantages they are unable to obtain with IBM peripherals. Amdahl Corporation is the first firm that has chosen to compete on the basis of a mainframe that is compatible with IBM peripherals.

Over the next decade, rapid technological changes may be expected to affect this industry. Satellite-linked data systems, time-sharing services, and new minicomputer developments are already having an enormous impact on the economics and patterns of computer usage. Additional changes in computer architecture, memory capacity, and processing techniques are expected at any time.

It is difficult to predict how the strengths of the various firms will evolve. Computer equipment tends to be subject to product generations, with innovative hardware developments rendering older systems obsolete. IBM is expected to introduce its first fourth-generation systems by 1980. For other firms to remain competitive, they will have to produce comparable pieces of equipment within a short period of time. In addition to requiring substantial commitments of R and D funds, continued active participation in this industry will require access to large amounts of capital to finance the leasing of computer equipment. Although IBM has about $1 billion in short-term liquidity, many computer manufacturers are burdened with debt obligations.

U.S. firms tend to dominate the world industry. IBM maintains the largest individual share of the computer market in the United States, Western Europe, and Japan. Control Data, through its joint ventures with other firms, sells more peripheral equipment in Europe than do any of the local manufacturers. Digital Equipment enjoys a 30 to 40 percent share of the minicomputer market in every major industrial country except France.

Firms in the U.S. computer industry suffer constraints similar to those present in the aircraft industry: shortage of capital for expensive R and D and protected foreign markets. To overcome these obstacles, firms in the U.S. computer industry have taken measures similar to those taken by U.S. aircraft firms. In exchange for access to foreign markets and an infusion of foreign government funds for developing new generations of technology, the U.S. computer firm is prepared to share its know-how. Small, new computer firms, such as the Amdahl Corporation, find such arrangements particularly compelling during recurring periods of venture capital shortages in the U.S. Having earned diminishing returns on their computer assets and opted to sell out altogether, some U.S. firms have found a scarcity of interested native firms and have been forced to sell their operations to foreign enterprises, thereby enhancing the competitive position of the foreign firms vis-a-vis the U.S. industry. Most of the purchasers of U.S. computer technology, whether acquired through a joint venture or acquisition, are highly absorptive and intent on becoming internationally competitive.

Trends in the World Computer Industry

Western Europe

As a long-term producer of electromechanical calculators, IBM established overseas manufacturing operations years before it started to mass-produce computers. Having these facilities already in place greatly expedited its expansion into the European computer market. In contrast to IBM, as late as the mid-sixties, firms such as General Electric attempted to export computers to Europe from their plants in the U.S. Other companies, such as RCA, attempted to participate in the growth of this market largely through licensing agreements with European companies. Eventually, IBM's major U.S. competitors either established turnkey facilities or acquired struggling manufacturers in Europe.

General Electric's 1964 purchase of majority ownership of Machines Bull in France and Olivetti's computer division in Italy provided it with a strong continental production base. By the late 1960s, GE's share of the European computer market was second only to that of IBM. When it sold its computer business to Honeywell in 1970, Honeywell assumed GE's market position. Honeywell currently has about 9 percent of the West German computer market, 10 percent of the British market, and 13 percent of the French market. Other U.S. computer companies that are important in Western Europe include Burroughs, Control Data, Digital Equipment, National Cash Register, and Sperry Rand.

The most significant European computer firms are International Computers Limited, which in 1974 had a 32 percent share of the British market; Siemens, which had a 16 percent share of the West German market; Campagnie Internationale pour l'Informatique (CII), with 8 percent of the French market; and SAAB, with only 7 percent of the Swedish market. ICL is expected to increase its European market share of small business computers due to its recent acquisition of Singer's international operations. IBM's hold on these markets, however, was substantial in 1974: in West Germany, its market share was 63 percent; in the United Kingdom, 42 percent; France, 62 percent; and Sweden, 58 percent. If individual European governments did not provide preferential treatment to their "national champion" computer manufacturers, IBM's share of these markets would probably be even greater. Currently, no European computer company is responsible for even 1 percent of total world computer installations.

One of the most interesting developments in the European computer industry was the formation in 1973 of Unidata, a consortium designed to mount a unified attack on IBM. Although the British firm ICL originally seemed interested in joining this venture, Unidata was eventually composed only of CII of France, Siemens of West Germany, and N.V. Phillips of the Netherlands.

Coordination within Unidata was not easy to accomplish. In 1975, the top management of CII publicly accused Siemens of inadequately promoting the French firm's products in West Germany. Perhaps for this reason, in May 1975, CII entered into an agreement to merge with Honeywell's European subsidiary, Honeywell-Bull. Once this merger is completed, CII-HB will be the largest computer company under European majority control. The chief executives of Siemens and Philips viewed this prospect somewhat differently; they denounced the proposed merger as an extension of American technological dominance. In fact, the Honeywell product line is expected to supplant that of CII. By the end of 1975, both Siemens and Philips had withdrawn from Unidata.

Eastern Europe and the Soviet Union

Although researchers in the Soviet Union began to experiment with computers in 1949, Soviet computer hardware is still not competitive with that produced by the leading manufacturers in the U.S., Western Europe, or Japan. Western experts generally believe that Soviet technology excels in a few software applications[3] and component technologies (such as advanced bubble memory technology).

In 1969, the Governments of six CMEA (Council for Mutual Economic Assistance) nations (Bulgaria, Czechoslovakia, East Germany, Hungary, Poland, and the Soviet Union) agreed to develop jointly a line of computers to be known as the RYAD. Although all six countries produce components for these computers, production of the largest machines is concentrated in Poland and the Soviet Union. RYAD computers appear to be patterned on the IBM 360, but they are generally reported to be less reliable. They are also said to be excessively prone to service and maintenance difficulties.

Rumania, which is not a participant in the RYAD consortium, has developed its own line of computers, known as the Felix. This development was aided by a licensing agreement with the French firm CII. Through a joint venture with Control Data, Rumania has also begun to assemble its own computer peripheral devices.

Exports of Western computer hardware and technology to socialist nations are generally subject to limitations imposed by the members of COCOM (Coordinating Committee of the Consultative Group of Nations). This organization is composed of Japan and all NATO nations except Iceland. Its regulations attempt to restrict exports that may contribute to the military capabilities of socialist nations. Usually these restrictions apply to systems and technologies that are significantly more advanced than those that are available in Eastern Europe, the Soviet Union, Cuba, and the People's Republic of China. Exceptions to the COCOM rules are usually approved only after much intergovernmental negotiation. Stringent safeguards are frequently placed on such exceptions. For example, COCOM may require that a Western representative visit a computer installation in a socialist nation at frequent intervals in order to determine that it is not being used for military purposes. U.S. firms periodically claim that they have lost sales to other nations' corporations because of liberalization in COCOM restrictions. In light of the growing number of exceptions that have been granted in recent years, these rules may soon be altered to sanction the sale of more advanced technologies and equipment.

Few U.S. computer companies are actively engaged in licensing arrangements with socialist partners. Western firms that currently maintain such arrangements include Control Data, ICL of Great Britain, Siemens of West Germany, CII of France, and the Japanese firm Fujitsu.

Much of the technology released to socialist nations is obsolete by Western standards. Some of it is embodied in turnkey plants. Some creates only a partial capability and necessitates further imports of Western components. For example, in the early 1970s, the U.S. firm Data Products transferred computer line printer technology to Videoton in Hungary. This technology transfer was sufficient for the production line printers except for the hammer banks, which still must be purchased from the U.S. company.

The People's Republic of China

Since the early 1960s, when the Soviets abruptly withdrew their technical assistance, the People's Republic of China has pursued a policy of developing its technological base without becoming permanently dependent on foreign suppliers. Although this approach does not preclude importing equipment or technology, the Chinese have essentially developed their own computer systems without help from abroad.

A visitor to China in 1972 gave the following report on its computer developments.[4] At the Institute of Computers of the Chinese Academy of Sciences, he examined a computer that had a memory speed only one-twentieth that of an IBM 370/145. Chinese engineers were engaged in a substantial research program on large-scale integrated circuitry, but they had not yet achieved breakthroughs comparable to those that had occurred in the U.S. by the late 1960s. Still, the visitor concluded that within a few years the Chinese would master this technology.

The PRC may soon receive some U.S. assistance in the development of computer technology. In April and May 1976, at least two U.S. manufacturers sent to Peking corporate executives who previously had negotiated licensing agreements with other socialist nations.

Japan

The development of the Japanese computer industry was aided by the erection of nontariff barriers. In 1957, the Diet passed the Provisional Electronic Promotion Laws, which prohibited foreign computer firms from establishing subsidiaries in Japan. By the end of the 1950s, local manufacturers were at a severe disadvantage because they had not gained access to the basic electronic data processing patents held by IBM. In 1961, in exchange for agreeing to disclose these patents to Japanese manufacturers, IBM was allowed to establish a wholly owned subsidiary in that country. Because imports of computers to Japan were greatly restricted, other U.S. manufacturers were unable to participate in the growth of this market except through licensing agreements. During the 1960s, Hitachi, Mitsubishi, Nippon Electric, Oki, and Toshiba licensed computer technology from RCA, TRW, Honeywell, Sperry Rand, and GE, respectively. Fujitsu was the only Japanese manufacturer not to license computer technology from abroad at this time. In 1972, it obtained Amdahl Corporation's fourth-generation computer technology, the most advanced currently incorporated in any mainframe.

The Ministry of International Trade and Industry has attempted to coordinate the development of the local computer industry. In addition to providing substantial subsidies for research and development, in 1971 it recommended a loose consolidation of the six Japanese computer firms into three government-supported producer associations, each specialized in a different portion of the industry.

Behind nontariff barriers, Japanese manufacturers have been able to develop a position of relative strength. MITI was therefore willing to lift most of these restrictions in late 1975. Foreign firms may now establish wholly owned subsidiaries in Japan and engage in unrestricted imports of computers. It is expected that U.S. companies will increase their penetration of the Japanese market, but the indigenous firms should be able to retain much of their position.

Recently, some Japanese companies have demonstrated a willingness to compete in foreign computer markets. A small number of minicomputers are currently being exported to the United States. Fujitsu's joint venture with the Spanish postal service and its attempt to exploit the Amdahl technology through a joint venture with Siemens are further evidence of the current state of Japanese computer technology.

Developing Countries

Outside of the industrially advanced countries and the socialist nations, there are few computer developments of any significance. One exception to this statement may be the joint venture Control Data is attempting to establish with the Iranian Electronics Industries Group. Although this joint venture will only produce computer terminals, it will be 70 percent owned by Iranian interests and is expected to achieve a substantial level of export sales to Western Europe. The joint venture partners hope that the Iranian plant will begin volume production in the early 1980s.

Another significant exception is the Brazilian state-owned computer firm, Cobra, which was incorporated in 1975 as the first of several anticipated national computer manufacturers. In early 1977, Cobra entered into a licensing and technology transfer arrangement with the U.S. computer firm, Sycor Inc. of Ann Arbor, Michigan, for the production of a minicomputer, to be designated the Argus 400. The Brazilian government is committed to establishing a national computer industry and will, in all likelihood, prevent foreign competition from affecting sales of the Argus 400. Such a decision will have an immediate impact on IBM's plans to introduce to the Brazilian market its newly developed minicomputer, the System 32.

Case Study: The Transfer of Amdahl Corporation Fourth-generation Computer Technology to Fujitsu

Many of the corporations that have attempted to confront IBM in the heart of its general-purpose computer business have ultimately been willing to settle for a comfortable market niche for the production of peripheral equipment or special purpose computers. Due to the competitive strength of this firm, the industry's leader and the nation's seventh largest corporation, several prosperous conglomerates have abandoned the industry—GE in 1970, RCA in 1971, and Xerox in 1975.

Amdahl Corporation was founded in 1970 with an initial capitalization of $33,000 and the intent to compete with IBM, whose sales for 1975 totaled $14.4 billion. Amdahl has been able to position its product against IBM's top-of-the-line mainframes (or central processing units) due to the technology and strategy developed by the firm's principal founder, Gene M. Amdahl. Substantial financial and production support for this effort has been provided by Fujitsu Limited, a Japanese diversified manufacturer of computers, telecommunications equipment, and electronic components whose sales totaled $765 million in 1974. The close relationship that has developed between these firms has led to joint development, cross-licensing, coproduction, and joint-venture sales agreements. As a result of this interaction, there has been considerable transfer of technology to Fujitsu, including the technology required to produce the world's most advanced computer—a machine that many experts consider the first of the fourth generation.

In late 1970, Amdahl left IBM, where he had headed the technical team that designed the 360 series of computers and directed the corporation's advanced computer systems laboratory. He carried with him a plan for challenging his former employer's preeminence in the most technologically advanced segment of the industry: the manufacture of large-scale, general-purpose mainframes. Previous attempts to weaken IBM's hold on this portion of the market had failed for three reasons: IBM's technology tended to be one step ahead of that of its competitors; it had substantially greater funds with which to pursue computer developments; and other firms' systems typically suffered from relatively disadvantageous packages of software and peripheral equipment. In its most rudimentary form, Amdahl's plan was to manufacture a mainframe that would be interface-compatible with IBM hardware and software; thus, in order to compete, Amdahl Corporation needed only to produce a central processing unit (cpu) that would provide a significant price-to-performance advantage over its IBM counterpart. Systems compatability would enable users of the IBM equipment to improve the cost effectiveness of their data processing by merely substituting cpu's.

Amdahl Corporation assumed that purchasers of its equipment would be able to continue to utilize IBM software, much of which was already in the public domain. IBM could institute changes that might frustrate Amdahl's plan to piggyback on IBM software and peripherals, but such behavior might provoke antitrust litigation because IBM had periodically been accused of following unfair practices designed to drive other firms from the industry. Amdahl operated under the assumption that IBM would not attempt to block its entry.

Designing an IBM systems-compatible mainframe would have been an almost impossible task for a team of computer specialists not intimately familiar with the design and production of this equipment. Gene Amdahl, however, had had appropriate experience. He was also in a good position to design a superior cpu. Before leaving IBM, he had conceptualized a method for upgrading the 360 to a level sufficiently advanced to constitute a next-generation computer. Third-generation computers utilized small-to-medium scale integration in their cpu logics and thus required between four and six circuits on every semiconductor chip, the basic feature that determines a computer's storage, processing, and retrieval capabilities. Unlike these machines, the Amdahl computer was designed to utilize large-scale integration (LSI), placing up to one hundred circuits on a chip. LSI technology was already in existence but had never been successfully harnessed for use in computers.

By using a computer-aided design (CAD) that Amdahl created after leaving IBM, a master photolithographic mask was developed for the production of semiconductor devices known individually as emitter-coupled logic (ECL). The CAD included software and design tools for developing the LSI circuits and determining the optimum component layout to maximize the speed, density, and flexibility of circuit location. The ECLs enable the Amdahl design to overcome the speed limitations imposed by the use of conventional bipolar integrated circuits and result in a much faster and more compact machine. Although Amdahl did not create ECL technology, his use of the master photolithographic mask was a conceptual breakthrough. Previously, ECLs were too costly and difficult to use. Various firms had unsuccessfully tried to overcome these problems by either standardizing or customizing ECL production. The mask allowed ECLs to be uniformly mass-produced up to a certain point and final-processed into 120 different parts.

The LSI chips that would be produced as a result of this technology would be mounted in chip carriers, about forty of which would be bonded to a ten-layer circuit board known as a multichip carrier (MCC). All bonding operations would be performed with customized processing equipment designed and built by Amdahl. About 85 percent of the required interconnections between the chip carriers would be incorporated in the printed circuit boards. Each MCC would be individually tested for continuity and performance. The final step would be to mount the MCCs in an LSI gate-frame assembly. Because of the large number of circuits incorporated in the MCCs, the computer would require only one-tenth the number of packaged semiconductor chips and would have fewer than one-third the number of solder terminations found in third-generation computers. The machine therefore would have one-third or fewer of the usual sources of operating failure. It would also be easier to install and maintain.

The Amdahl cpu was designed to provide several advantages over its IBM counterparts. Because of its LSI technology, the new cpu would enable users to realize cost savings of up to 50 percent and would occupy one-third as much floor space. The cpu would be air-cooled rather than water-cooled, and this together with other features would make it much easier to maintain. Each machine would also contain a diagnostic control console that would pinpoint the source of any major difficulty. This terminal would also be able to interface with Amdahl's diagnostic center in Sunnyvale, California, should any unusual problems arise.

The different stages of the development may be described as design, simulation, prototype development, and production-engineering. The design stage would consist of the conceptualization of the organization, logic, and electrical-mechanical aspects of the computer. In the simulation stage, all of the hardware except the fourth-generation LSI logic would be tested. The exact specifications of the advanced logic would be determined during this phase. The building and final testing of the completed prototype would be the subject of the next stage. Finally, production-engineering would lead to mass production.

Amdahl planned to market its first computer in 1973. It hoped to achieve between 10 and 15 percent penetration of a total market not to exceed 1000 units by the end of 1977. Amdahl realized that the introduction of new IBM products could render its product line obsolete at any time. New computer product generations had historically occurred every five to seven years. Given the timing of the introduction of IBM's most recent models, Amdahl doubted that its product line would become obsolete in the near future. In the meantime, it planned to take sales away from IBM and to upgrade its own product.

Raising capital from U.S. sources for this venture was a difficult task. During Amdahl's first two years of existence, only $12.8 million was raised from U.S. citizens and institutions, including $5 million from Heizer Corporation, a venture capital firm. Other principal investors in Amdahl included Nixdorf Computers, a West German firm that purchased $6 million of common stock, and Fujitsu Limited, which invested $6.2 million in stock and $5 million in interest-yielding notes. Nixdorf and Fujitsu also placed orders for Amdahl computers, so the fledgling manufacturer had a $35 million backlog of orders.

For several reasons, Fujitsu was unique among Japanese computer firms. During the 1960s, Hitachi, Mitsubishi, Nippon Electric, Oki, and Toshiba had licensed technology from RCA, TRW, Honeywell, Univac, and GE, respectively. Fujitsu was the only Japanese computer manufacturer that had not licensed technology from abroad. It was also the only one to earn over half its revenues from computers, to engage in substantial export sales, and to operate profitably. These accomplishments should be viewed in relation to the development of the Japanese industry.

In 1957, the Diet passed the Provisional Electronic Promotion Laws, which assigned the task of guiding the development of the local computer industry to the Ministry of International Trade and Industry (MITI) and restricted the ability of foreign firms to manufacture in Japan. Prior to 1960, only National Cash Register was allowed to produce computers in the country. This exception was due to a "grandfather clause" that provided special treatment to foreign firms that had production facilities in Japan prior to World War II. By the end of the 1950s, Japanese manufacturers were at a severe disadvantage because of their lack of access to the basic electronic data processing patents held by IBM. In 1961, in exchange for agreeing to release these patents to local manufacturers, IBM was allowed to establish a wholly owned subsidiary in Japan. No other foreign firm had a similar advantage to relinquish, so none of the others was afforded equal treatment. Still, IBM-Japan was not considered a Japanese company: its products were viewed as imports, even if locally produced, and their purchase by Japanese users had to be justified to MITI on a case-by-case basis. Actual computer-products imports could occur under the provisional laws, but these were subject to quotas and tariffs, which in 1971 were raised from 15 to 25 percent. Given these circumstances, important U.S. computer manufacturers agreed to license their technology as it obsolesced. In this way, the Japanese acquired much of their second- and some third-generation technology. By the late 1960s, Japanese manufacturers were still weak in their production of peripheral equipment, software, and large memory hardware.

Behind these barriers, import restrictions, and constraints against foreign ownership of local manufacturing, Fujitsu developed a position of relative strength. By the end of the decade, its share of the home computer market surpassed that of any other Japanese firm. Yet in comparison with IBM, the dominant firm in this market, Fujitsu suffered from several deficiencies. It had particular difficulty producing large computers and adequate supplies of advanced components. Fujitsu's software also needed improvement, although this shortcoming was not critical due to the routine applications Japanese users typically found for their computers. Still, Fujitsu was technically quite competent and able to satisfy the needs of much of the home market.

Perhaps because of its market position and its penchant for independent development of its computer technology, Fujitsu tended to be favored by MITI. Starting in 1970, MITI subsidized Fujitsu's R and D on fourth-generation computers. It is believed that by 1978, MITI will have provided close to $100 million for this purpose.

In October 1971, when MITI recommended channeling the six Japanese computer firms into three government-sanctioned producer associations, each specializing in a different portion of the industry, MITI displayed further evidence of favoritism. It selected Hitachi as Fujitsu's partner. Hitachi was the largest and most diversified Japanese company producing computers. As such, it had vast technical, financial, and managerial resources on which it could draw. In computer sales it trailed only IBM and Fujitsu.

At approximately the same time as the Hitachi-Fujitsu relationship was being formed, Fujitsu entered into its first technology-sharing arrangement with Amdahl. The two companies agreed to develop jointly a virtual storage memory[5] for the Amdahl machine then on the drawing board. At that time, Amdahl was planning to produce two versions of its fourth-generation computer—a "real memory" model and a system with virtual memory. Although a spokesman for Fujitsu later asserted that its aid led to Amdahl's virtual storage design,[6] information supplied by Amdahl in 1973 to the Securities and Exchange Commission conveys a different impression. According to this source, the two firms' engineers worked closely together on a day-to-day basis in the developmental stages of engineering and manufacturing this model.[7] The implication is that the design was solely Amdahl's but that the two firms jointly performed the production-engineering. This was but the first of several instances in which the two firms shared technology.

The next occasion on which Amdahl and Fujitsu interacted was in February 1972, when the latter made a substantial investment in Amdahl. Between that date and November of the same year, Fujitsu California, Inc., a subsidiary of Fujitsu Limited, purchased $5 million worth of Amdahl convertible debentures. It bought $6.2 million of common stock in December. Perhaps not coincidentally, on 29 December 1972 the two firms entered into an exclusive, royalty-free cross-licensing agreement for their respective patents and know-how relating to the company's products that were then under development.[8] Although Amdahl's S-1 statement to the SEC is rather vague, subsequent journalistic accounts have identified what was transferred to Fujitsu—the ability to incorporate LSI technology in its computers.[9] It is probable that in exchange for transferring the ECL-LSI technology, Amdahl received both financing and the use of some patents of far less value than those it transferred. In any case, in terms of the flow of unique technology, this transfer was decidedly one-directional. It is evident, however, that as a result of this cross-licensing agreement, there was a strong flow of production-engineering and technical support to Amdahl. At least twenty-five Fujitsu engineers were sent to California to aid the U.S. firm's technical team.

Under the terms of the cross-licensing agreement, Fujitsu was granted exclusive patents for Japan and Amdahl obtained patents for the U.S. On 14 April 1973, the two firms signed a memorandum of understanding indicating their intention to form a vertically integrated, jointly owned company to develop, manufacture, and sell computer products in areas of the world where the two companies have nonexclusive patent rights.[10] On 9 October 1974, these agreements were modified to grant Fujitsu additional exclusive patent rights for Spain and Amdahl took patent rights for Canada.

It is interesting to note that two items were definitely not transferred to Fujitsu—Amdahl's computer-aided design (CAD), which was necessary for the production of the master photolithographic mask for designing the ECLs, and Amdahl's final test equipment. The retention of these items provided Amdahl with a measure of leverage over Fujitsu. Without the CAD, the Japanese firm would have great difficulty leapfrogging the technology it received from Amdahl; that is, it would find it very difficult to develop a new mask so it could produce upgraded ECLs. Of course, it could develop its own CAD or acquire this technology from Amdahl. In neither case were its efforts certain to succeed. By not transferring the know-how for the final test equipment, Amdahl preserved a reliability advantage for its computer. By holding exclusive patents to its technology for the U.S. and by retaining control over vital portions of its core technology, Amdahl preserved the option of much future independence. It alone could exploit its current technology in the U.S., and with an upgrade it could become completely free of Fujitsu.

Design changes at IBM and problems in cash flow, product planning, and management prevented Amdahl from meeting its desired schedule. The company had leased two factory buildings in Santa Clara, California, and had hired more than 300 people. But in mid-1973, IBM introduced its virtual memory 370-series computers, thereby making obsolete Amdahl's real memory model, which Amdahl had hoped shortly to put into production. As Amdahl was without revenue and needed an additional $20 to $30 million in order to manufacture virtual memory computers, it tried to raise capital through a public stock offering. The timing for such an effort was poor: the stock market was depressed; venture capital was becoming extremely difficult to attract; and the memory of GE and RCA withdrawing from competition with IBM was fresh in the minds of potential investors. When this offering failed, a management shakeup occurred. Gene Amdahl was allowed to retain his position as chairman of the board of directors, but a new chief executive officer was brought in to manage the company.

The new management convinced Fujitsu and Heizer to double their investments, thereby providing an additional $17 million to Amdahl and increasing Fujitsu's ownership of the company to approximately 41 percent. Under the Credit Facilities Agreement signed by both firms on 3 April 1974, Fujitsu purchased an additional $7.14 million in convertible notes and a part of Amdahl's inventory for $4.5 million. At the same time, Amdahl agreed to grant Fujitsu nonexclusive, royalty-free, and irrevocable licenses under certain patents to manufacture, use, and market within the U.S. products covered by such Amdahl patents.[11] In short, in order to obtain financing, Amdahl was forced to surrender a significant advantage to Fujitsu. Presumably, this agreement does not apply to any Fujitsu products that might compete with Amdahl's product line in the U.S. There is other evidence to suggest that Fujitsu intends to sell in the U.S. smaller-scale computers that will incorporate the LSI technology gained from Amdahl.[12]

In addition to the credit agreement, the two firms also signed a manufacture and purchase agreement on 3 April 1974. Under this agreement, 80 percent of the Amdahl manufacturing effort was shifted to Fujitsu. The Japanese firm's scope of manufacture would extend to the unit testing of certain subsystems. These subassemblies would be produced in Japan with ECLs imported initially from the U.S. and later produced in Japan by Hitachi. The subassemblies would be shipped to California, retested with Amdahl's own test equipment, and assembled with consoles, power units, and other items that Amdahl would either manufacture or purchase from other sources. Meanwhile, Fujitsu would be free to utilize subassemblies of these same specifications in its own Amdahl-like computers. In order to implement this agreement, Amdahl agreed to furnish Fujitsu a complete engineering documentation package for the system.[13] This package would consist of all engineering drawings and specifications required to describe the technical configuration of the system, whether classified as released or prereleased. As evidenced by the level of detail required in this transfer of know-how, the shifting of the manufacturing effort led to a more complete and more easily absorbed transfer of technology than would have occurred under just the cross-licensing agreement. From Amdahl's perspective, having its computers produced by a manufacturer already in place was clearly preferable to establishing one on its own. Transferring the necessary technology as effectively as possible was now very central to Amdahl's self-interest.

Under a manufacturing agreement dated 11 October 1974, Amdahl agreed to purchase between 42 and 108 systems between January 1975 and May 1977. It may purchase up to an additional 108 systems between June 1977 and May 1979. The price for these units is fixed, subject to adjustment to fluctuation in the electrical machinery section of the U.S. wholesale price index. If Amdahl fails to purchase the minimum number of systems or if Fujitsu is unable to deliver, a schedule of penalties will be imposed. Also, if Amdahl does not pay any such penalties within two months of their imposition, Fujitsu may market and deliver a sufficient number of systems in the U.S. such that at least 42 are installed.

In the summer of 1975, both Amdahl and Fujitsu signed their first agreements to install fourth-generation computers. The first installation of the Amdahl machine, which has been designated the 470 V/6 and which is functionally equivalent to an IBM 370/168, occurred in June at NASA's Institute for Space Studies in New York. At least six other Amdahl machines have subsequently been installed and at least twenty-seven letters of intent have been signed. Fujitsu's first sale of its 470-equivalent, the M-190, was made to the Spanish national telephone company. Fujitsu and the Spanish firm have also established a joint venture for the production of computers in Spain.

The initial performance of the 470 V/6 has been excellent. The changeover from NASA's IBM 370/165 took less than two days to accomplish. The new machine provided an immediate 30 to 40 percent improvement in computational speed. Amdahl hopes that the 470 V/6 will prove to be two and a half times as fast as the 165 under appropriate conditions. To achieve a similar level of performance, NASA would have had to purchase a 370/195. The sales price for such a machine would be roughly double the $3.8 million cost of the Amdahl computer. While a 195 could not have fit into the NASA computer facility, the 470 V/6 is small enough to fit into a large closet.

Before assessing the competitive impact of this transfer of technology to Fujitsu, it is necessary to consider additional information about both the

Japanese computer industry and IBM. Certain facts from the previous discussion also bear repetition: Without Amdahl's computer-aided design, Fujitsu will have great difficulty upgrading the ECL technology; the Japanese firm is apparently allowed to market small-to-medium-scale computers in the U.S.; and IBM can outmode Amdahl's technology by placing a significantly more advanced computer on the market.

During the late 1960s, Japanese computer firms presented little competition to U.S. manufacturers. Not only was their technology largely imported from the U.S., but in many cases it was also either obsolete or handicapped by the absence of other technical developments. For these reasons, in order to be viable, Japanese computer firms had to operate behind tariff and nontariff barriers. However, with each passing year, the Japanese industry has gained additional strength, and the barriers therefore have diminished. This liberalization was completed in 1975, at which time computer import and capital investment restrictions were completely removed. Foreign firms may now purchase equity and even achieve majority ownership of Japanese computer companies. The Japanese have been willing to accept this liberalization because of the great strides their computer firms have made in recent years. Many of these accomplishments have been facilitated by the Ministry of International Trade and Industry's infusion of $300 million in subsidies. Sixty-five million dollars has been allocated for R and D on very large-scale integrated circuitry (VSLI), a technique that, if successful, will enable manufacturers to place up to 1000 circuits on a semiconductor chip. If the Japanese achieve this breakthrough, their firms may compete vigorously with the U.S. industry.

It should be noted that Japanese firms currently provide only 3 percent of the world's annual computer production, whereas U.S. manufacturers produce approximately 70 percent. Whether the relationship between these market shares will more or less remain fixed depends on many factors, including IBM's ability to retain its market position.

According to the trade press, IBM is currently planning to market its own series of fourth-generation computers. The conventional wisdom is that IBM could put these machines on the market very shortly but that it chooses not to do so in order to increase returns on its current product line. If Amdahl and Fujitsu succeed in making major inroads against IBM, this circumstance would almost definitely influence IBM's timing.

IBM's next generation has been named the "Future System." This system is designed to utilize special software that will facilitate communication between the user and the computer and thus reduce the importance of the computer programmer. Descriptions of this system indicate that it is designed to appeal primarily to users with sophisticated programming needs. Thus, these computers may be designed primarily for the North American and Western European markets.

Most probably, Amdahl and Fujitsu will successfully compete against IBM for the next two to three years. Each firm by itself will sell large-scale ECL-LSI computers in those markets where it holds exclusive patent rights. Sales to all other markets will be on a joint venture basis. Additionally, Fujitsu will market smaller-scale computers in Japan, the U.S., Spain, and elsewhere. Whether the Future System will render large-scale ECL-LSI computers obsolete is not clear, although one must assume that IBM intends to deal effectively with its high-technology competitors.Certainly, it will be more difficult for Amdahl to develop a machine compatible with the Future System than it was to compete with the 360 and 370 series computers. To the extent that Amdahl and Fujitsu are able to develop computers technologically more advanced than the Future System, they will develop long-term staying power. Alternatively, each firm may have as its ultimate goal dominance of a particular segment of IBM's current market or product line. With further consolidation of the Japanese computer industry, additional subsidies, and important technological breakthroughs, Fujitsu may be able to do more than just influence IBM's timing; it may take away part of its market share. It has been estimated that the infusion of Amdahl's technology has allowed Fujitsu to close approximately a three-to-five year technological gap between it and the U.S. industry. By 1980, the Amdahl-Fujitsu joint venture may sell a sufficiently large number of its advanced systems in North America, Western Europe, and Japan to displace more than $500 million in revenue to IBM.

Case Study: Honeywell's Transfer of Technology to Toshiba, Nippon Electric, and CII-HB

In 1974, Honeywell, Inc. was the nation's sixty-eighth largest corporation, with over $2.6 billion in sales. It operated wholly owned subsidiaries in many countries, including Australia, Canada, France, Mexico, South Africa, the United Kingdom, and Venezuela. In terms of worldwide computer revenues, Honeywell was second only to IBM ($1.3 billion versus $9.9 billion). Its world market share, however, was less than 10 percent, while IBM's was approximately seven times as large. It was the majority shareholder of Honeywell-Bull, a major computer operation headquartered in France. Additionally, it participated in a joint venture in Japan, Yamatake-Honeywell Company, Ltd. The latter held licenses for much of the firm's noninformation systems technology. Honeywell's computer technology was licensed in Japan to Toshiba and Nippon Electric Company (NEC).

Toshiba and Nippon Electric

In October 1974, Honeywell purchased the vast majority of General Electric's computer manufacturing assets. GE initially retained its time-sharing services and industrial process control computers, but sold the latter to Honeywell in 1974. The firm simply could not afford to continue to undertake the enormously expensive R and D efforts required to remain competitive in the computer industry.[14] GE was already deeply engaged in equally expensive projects—aircraft engines and nuclear power plants.

As a result of the purchase, Honeywell inherited a majority-owned computer manufacturing subsidiary in France and a licensing agreement with Toshiba.[15] Since Honeywell and GE had licensed computer technology to their Japanese partners on somewhat different terms, Honeywell moved rapidly to bring these agreements into conformity. Royalty rates were equalized and each firm was afforded equivalent access to the U.S. company's technology. Honeywell authorized Toshiba and Nippon Electric to sublicense each other. When, in 1971, the Japanese Ministry of International Trade and Industry (MITI) suggested that the two join forces in a domestic computer cartel, Toshiba and NEC were brought even closer together. Today, the two firms coordinate research and product line developments.

In December 1975, MITI lifted the nontariff barriers that previously prevented foreign firms from establishing computer subsidiaries in Japan. Shortly thereafter, Burroughs Corporation announced its intention to take over its Japanese affiliate, which was a particularly weak force in the local computer industry. To date, Honeywell has given no indication that it plans to alter its relationship with its Japanese licensees.

Honeywell experienced little reverse flow of technology from Toshiba and NEC until 1975. These firms now have begun to provide a meaningful reverse flow, particularly of technology related to computer peripheral equipment. The Japanese government is currently subsidizing Toshiba's and NEC's research and development on very large-scale integrated-circuit computers. Because of their cross-licensing agreements with Honeywell, should one of these firms achieve a substantial breakthrough, the U.S. firm would gain access to information on the new semiconductor chip capabilities. It would not, however, automatically obtain manufacturing techniques, circuit design, or logic characteristics. On the other hand, if a breakthrough occurs in which Honeywell technology is upgraded, this information would be automatically provided to the companies.

CII-HB

Honeywell's relationship with CII-HB is more difficult to describe than the rather conventional licensing arrangements just discussed. The 1975 agreement to merge Honeywell-Bull into Compagnie Internationale pour l'Informatique (CII) contains several themes of interest. In large measure, this merger climaxes fifteen years of technology transfer to the French computer industry.

In 1931, a Frenchman named Georges Vieillard paid $4000 for patents on a primitive data processing machine that utilized wooden punch cards. With this technology in hand, he founded Machines Bull. In 1952, the company introduced the world's first germanium diode computer. For several years thereafter, Bull's products maintained a technological edge over IBM's. During the late 1950s, the Bull Gamma 3 computer not only matched U.S. machines in quality, but it also sold for less.[16] Although Machines Bull did not export computers to the U.S. during this period, it sold input equipment to Remington Rand, Univac, and other U.S. manufacturers for use in their computers. By 1959, Machines Bull had a 40 percent share of the European computer market. IBM's share at this time was 50 percent.

In 1960, IBM introduced second-generation computers that employed transistors and circuit boards in place of vacuum tubes and modules for circuit construction. Because Machines Bull was unable to match this technology, in 1961 it signed a ten-year licensing agreement with RCA. The latter's 3301 computer was designated the Bull Gamma 40 and was rapidly put into production in France to compete with the new IBM machines. Due to the financial terms of this license, however, Bull made little profit on these sales. One year after the introduction of the Gamma 40, Bull's revenue rose by 21 percent and its profits fell by 75 percent. Skyrocketing inventory costs and other management problems contributed to this erosion.

In 1962, General Electric exported computers to Europe from its Phoenix, Arizona, plant. In order to establish a European production and marketing base, it offered to purchase a 20 percent share of Machines Bull. When the French Minister of Finance, Valery Giscard d'Estaing, took GE's offer to President de Gaulle, it was rejected.[17] A French group then tried to refinance Machines Bull, but this effort failed. As a result of Bull's inability to regain its previous strength, in 1964 the French government decided to allow GE to acquire a 50 percent interest in the company and to obtain management control.

The Johnson administration stimulated the next major development in the French computer industry. In 1966, it prohibited the sale of U.S. advanced computers for use in the French nuclear and space programs. This slap at the French *force de frappe* led directly to the formation of Compagnie Internationale pour l'Informatique and to recognition of it as a stepchild of the state. The next national *plan calcul* provided substantial subsidies and preferential treatment for French computer firms. Because Machines Bull was under the control of GE, it was ineligible for these benefits and was largely excluded from selling computers to the government.

By 1967, GE had poured more than $200 million into its European computer subsidiary and had committed several tactical errors.[18] The top executives chosen to run GE-Bull lacked basic experience in both computers and European business practices. Not only did they arbitrarily impose GE management methods, but they also replaced the Bull product line with GE models that were only slightly more advanced. Furthermore, the GE managers decided to close one of Bull's five plants and considered additional measures to reduce the number of employees on the payroll. French workers enjoy more job security than their American counterparts, and the French GE workers, managers, and technicians were highly antagonized by this action. Despite these problems, however, GE eventually obtained a stronger position in the European computer market than it enjoyed in the U.S. In 1970, this strength passed to Honeywell when it acquired the majority of GE's computer holdings.

Between 1966 and 1975, CII received more than half a billion dollars in French government subsidies. Despite preferential treatment, it continued to operate at a loss. Competition with IBM, Honeywell-Bull, and others was particularly keen. In an effort to become more competitive, it spent millions of dollars on research, but many observers were skeptical that CII would become competitive in the near term.

In 1973, CII decided to join forces with Siemens of West Germany and N.V. Phillips of Holland in a European computer consortium known as Unidata. These firms agreed to coordinate their efforts in order to mount a unified attack on IBM. By 1975, CII top management concluded that Siemens' promotion of the CII product line in West Germany lacked conviction. The French Minister for Industry and Research was particularly resentful of Siemens's takeover of the computer operations at AEG-Telefunken, which competed directly with CII in the sale of large systems.

In 1975, articles in the U.S. trade press intimated that Honeywell might shortly abandon its computer operations.[19] Some analysts may have questioned Honeywell's ability to match IBM's developments on a timely basis. The previous year, IBM retained close to a billion dollars for R and D, while Honeywell raised only $305 million for development on a wider range of products.

Although Honeywell introduced a new family of computers in 1974, at least one commentator felt that the company was losing its technological strength, which had come, at least in part, from GE.[20] This criticism aside, the new systems gained broad acceptance with users. Unfortunately for Honeywell, due to the high interest rates then prevalent, most customers preferred rentals to outright purchases. Rentals strained the company's cash resources and postponed the realization of profits for several years, but the situation did not lead to unmanageable cash flow problems.

In March 1975, Honeywell and the French government announced a plan that was expected to help both parties. CII and Honeywell ratified this proposal on 29 December 1975. In mid-1976, for approximately $58 million, Honeywell sold 19 percent of its European subsidiary to French interests. This transaction reduced Honeywell ownership of its affiliate to 47 percent but allowed French interests to hold the remainder. Because Honeywell-Bull was then a majority French company, it was permitted to merge with CII. Although Honeywell does not have management control over the new entity, its representatives constitute a majority of the CII-HB technical committee, the mechanism that controls the firm's technology and product development. In order to assure a steady flow of Honeywell technology, it is expected that CII-HB top management will pursue a marketing strategy compatible with that of Honeywell's.

As a member of CII-HB, Honeywell expects to receive $280 million in French government subsidies over the next four years. (Presumably, technology developed by CII-HB will be available to Honeywell on a royalty-free basis by right of cross-licenses.) Due to "buy French" procurement policies, CII-HB will sell computers to the French government. The value of these sales for Honeywell is expected to reach $400 to $500 million over the next four years. These sales will provide the U.S. company with a substantial cash flow and immediate profits.

In terms of annual sales revenues, CII-HB will be as large as Honeywell Information Systems' U.S. operation. Some analysts believe in the so-called "10 percent rule" according to which, if a firm's share of the world computer market is above 10 percent, it may expect to earn at least 10 percent profit on sales; if its market share is lower, its profits are likely to be negligible. Should such an economy of scale actually materialize, the CII-HB merger might greatly improve Honeywell's profitability.

From the French government's perspective, the CII-HB merger is equally attractive. CII has received up to $100 million a year in public subsidies. Because of the merger, these payments are expected to cease in four years. In addition, the French have obtained "ownership" of Honeywell-Bull and access to an ongoing flow of U.S. advanced computer technology. French national security interests will be safeguarded by keeping CII's manufacture of military application computers separate from CII-HB operations. Together with minicomputers, such machines will continue to be produced in CII, a part owner of CII-HB, which will remain independent of foreign equity.

Siemens and Phillips were upset by the announcement of the proposed merger, which they perceived to subordinate European to French national interests. Perhaps for this reason, before the end of 1975, both firms withdrew from Unidata. For at least several years to come, CII-HB will be IBM's largest competitor in Europe.

For the first few years after the initiation of the licensing agreements with General Electric and Honeywell, Toshiba and NEC purchased large quantities of components and production equipment from the U.S. As the

Japanese firms' capabilities increased, the imported content of their computer products diminished. At present, the dollar value of their purchases from Honeywell are approximately equal to their royalty payments to the U.S. firm. Neither Toshiba nor NEC engages in significant exports of computer products.

In contrast, Honeywell-Bull's production has been integrated into the parent company's international marketing and production system. At least one of the models in the new System 60 family of computers is produced only by Honeywell-Bull. When Honeywell leases or sells one of the HB machines, it is produced in France and shipped to the customer. Similarly, HB markets the total range of Honeywell computer products. This arrangement will continue after the merger has been completed, even though CII-HB will be a noncontrolled foreign affiliate.

Case Study: Control Data's Transfer of Computer Peripheral Technology Through International Joint Ventures

Control Data Corporation (CDC), the world's second largest manufacturer of computer peripheral devices, has entered into a variety of joint ventures since the early 1970s. From these joint ventures has emerged an agreement between CDC and other manufacturers to standardize much of the peripheral equipment their computer mainframes require. By encouraging several firms to procure the same models of equipment, these ventures are able to provide their parent corporations with substantial cost savings which result from economies of scale due to large volume production and the sharing of research and development expenditures. At least in part, CDC top management selected this strategy in order to compete with IBM, whose sales of computer products are approximately nine times as large ($9.9 billion versus $1.1 billion in 1974). If CDC had not taken this approach, its computer peripheral products might gradually have lost out in the cost competition to IBM.

CDC manufactures a broad line of computers, ranging from its very large STAR-100 and CYBER 170 mainframes to its System 17, a small-scale, general-purpose computer. In 1975, the company's computer business—data services, systems, and peripherals—accounted for $1.25 billion in revenues. Over $215 million resulted from the company's international operations. At the same time, approximately $49 million was expended on research and development efforts.

Companies such as International Computers Limited (ICL) of the United Kingdom and Siemens Corporation of West Germany rely almost exclusively on CDC peripherals. United States manufacturers such as Honeywell and National Cash Register (NCR) also extensively utilize CDC products. The company also manufactures memory devices for IBM computers and markets these items directly to that firm's customers, and produces a broad line of computer supplies, including disk packs, forms, punched cards, and magnetic tape.

Multinational Data

CDC's first experience with joint ventures began in 1971, when together with ICL and NCR it established Multinational Data. The purpose of this entity was to establish common computer standards, languages, and hardware. After about a year of working together, the parent organizations concluded that a common computer architecture would be infeasible for the immediate future; each company was too dependent on its own timing and approach to allow for such intensive joint development. Multinational Data did, however, succeed in establishing some common interface standards, component standardization, and exchange of technical information that has enhanced each firm's product line.

Multinational Data gradually became a loose association which was renamed STACK. This organization has continued to work for greater standardization and has led to joint procurement of semiconductor devices for its member corporations. By procuring standardized subcomponents through a jointly-owned purchasing company, CDC, NCR, and ICL have been able to obtain significant economies of scale. In the process, a limited amount of technical information has also been exchanged.

Computer Peripherals, Inc.

In 1972, CDC and NCR established Computer Peripherals, Inc. (CPI), a 50-50 joint venture to manufacture card readers, printers and tape drives. Although CPI sells products only to its parent organizations, Control Data has continued to sell comparable pieces of equipment to other original equipment manufacturers.

Essentially, each parent firm retains ownership of whatever patents it brings to the joint venture. Patents which result from jointly funded R and D belong, however, to the new entity. Their release to the parent corporations may occur only under terms and conditions to be set by the board of directors of the joint venture.

Late in 1975, the British firm ICL purchased one-sixth ownership of CPI. It is expected that during the next year, it will become a full partner in the joint venture.

ROM Control Data SRL

In April 1973, Control Data became the first U.S. corporation to establish a joint venture with a Rumanian enterprise. In early stages of the negotiations, Rumania's Industrial Center for Electronics and Vacuum Technology (CIETV) asked for a patent and know-how license for such peripheral equipment as printers, card readers, tape drives, and disk storage drives. CDC, however, wanted part ownership of a local production facility in exchange for the provision of its production knowledge. The final contract called for the creation of ROM Control Data SRL to manufacture a series of slow-speed line printers, card readers, and punches—55 percent of which would be owned by CIETV and 45 percent by CDC. The former partner provided $2.2 million in capital, and CDC contributed $1 million worth of equipment, and technology, documentation, and training valued at $800,000. The life of the agreement is 20 years.

CDC's transfer of technology had three components: 1. technical documentation (e.g., drawings and schematics), 2. production techniques (e.g., flow soldering and assembly methods), and 3. quality control and monitoring procedures. While these technologies are the most up-to-date for the specific models of equipment to be produced, they are not necessarily representative of state-of-the-art for these specific categories of products.

CDC was motivated by the profit potential of the joint venture and, therefore, rejected the other nonequity types of options proposed by the Rumanians. To help insure revenues greater than an annual technical-assistance fee might yield, they successfully negotiated a Rumanian state guarantee that ROM Control Data would realize a minimum pre-tax profit of ten percent. Estimates for its first five years of operation indicate that the joint venture will sell approximately $10 million worth of products, with annual sales thereafter expected to be about $5 million. CDC has figured that its projected guaranteed profits are several times what it would have earned by selling its technology through a technical assistance agreement.

In addition to earning larger profits, the joint venture was attractive to CDC for its service as a low-cost manufacturing source from which to supply Western European markets. During the first five years of the joint venture's existence, CDC is committed to purchase up to 80 percent of its annual output. After that period of time, as Rumania's needs and exports increase, CDC's commitment will drop to 45 percent. The peripheral equipment produced by ROM Control Data will meet the particular standards of several CDC European customers and will relieve the firm of customized manufacturing in its U.S. facilities. This releases production capacity for the U.S. market and results in lower unit costs.

A final but quite substantial advantage CDC gains from the joint venture is expanded access to the Rumanian market. Since the late 1960s, the U.S. firm had been but one of several foreign firms to sell computers and peripheral equipment to Rumania. Now, in addition to enjoying a monopoly position for at least three of its peripherals, CDC is hopeful that it can enter the "inside track" of the country's computer market. By purchasing most of the joint venture's initial output, CDC expects to increase the amount of hard currency available to Rumania for the purchase of additional computers and components. It is also conceivable that CDC can increase its sales from the joint venture by expanding the range of products produced.

Magnetic Peripherals, Inc.

In 1975, a 70-30 joint venture known as Magnetic Peripherals, Inc., was founded by Control Data and Honeywell, a large purchaser of CDC magnetic peripheral products. Although there is no foreign participation in this enterprise, a brief discussion of it is relevant to this study for the following reason. Due to the market position of Magnetic Peripherals and Computer Peripherals, Inc., they represent the largest producers of disks, magnetic tapes, punched card equipment and printers in the world. As such, they stimulate much de facto standardization of computer equipment in the United States and Western Europe.

Computer Terminals, Inc.

In December 1975, Control Data and the Iranian Electronic Industries Group signed an agreement to establish a joint venture to be named Computer Terminals, Inc. Should the U.S. export administration authorities approve the necessary transfer of technology, CTI will produce terminals for both the Iranian and Western European markets. Because the Iranian group will provide the vast majority of the joint venture's capital, it will own seventy percent of CTI.

Included in the new venture's announced plans is the development of a plasma display unit and a voice input terminal. Most of the initial research for these efforts will take place in the United States. It is not known whether CDC or CTI will retain the patent rights. In any case, production at the new firm is expected to begin in 1980. Iranian officials estimate that the new venture will generate $17 million in sales in its first year of operation. According to one source, CTI will be Control Data's most extensive transfer of technology to a foreign government enterprise.[21]

The CTI product line is not expected to overlap with that of ROM

Control Data, Magnetic Peripherals, or Computer Peripherals, Inc. As in these other cases, however, CDC may be expected to achieve significant benefits such as the sharing of R and D expenditures, lower unit costs, and a savings of investment funds for additional plant capacity.

Control Dataset, Limited

In March 1976, Control Data and International Computers Limited signed an agreement in principle to form a joint company to produce magnetic disks and tapes, punched cards, printer ribbons, and computer stationery and furniture. The new company will be known as Control Dataset, Ltd., and will be 75 percent owned by CDC and 25 percent owned by ICL. Control Dataset will take over the business products operations of CDC in the United Kingdom, with the exception of its facility in Brynmawr, Wales. Additionally, Dataset, Ltd., a subsidiary of ICL, will be subsumed into the joint venture. The new company is expected to be initially capitalized at $5 million.

Additional Negotiations

In various trade publications, it has been reported that Control Data is currently attempting to negotiate a joint venture with Japan's Nippon Electric Company and a technology swap arrangement with the Soviet Union. So far neither project has come to fruition. In the case of the Soviet Union, CDC would like to swap its 100-megabit disk technology for Soviet know-how of comparable value, perhaps that nation's advanced bubble memory technology. Whether such an arrangement may be feasible will depend on the willingness of the Soviet Union and the approval of the U.S. Commerce Department's Office of Export Administration.

In deciding to participate in a transaction which will lead to a transfer of technology, Control Data Corporation considers: 1. the existence and availability of alternate sources of technology, and 2. the stage of technological development of the recipient. Ideally, if the buyer can procure the desired product and/or technology from other sources, CDC can discount most of the negative implications of its making the sale itself (e.g., impact on the company's level of employment, impact on the U.S. economy, etc.). The level of the recipient's technological sophistication is relevant inasmuch as it affects the seller/buyer relationship, use of technology and market. The resources and needs of the buyer may be such that its market is limited to equipment purchases and/or leases. On the other hand, the purchase may be of technology and be motivated by a desire to develop a "boot straps" capability so that it can be relatively independent in its future needs or become a competitor in certain markets.

Because CDC has been faced with increased competition, market maintenance and market entry problems, it has been encouraged recently to enter into relationships which involve a transfer of technology greater than that which normally takes place in the sale of equipment. This most often involves a joint venture. It offers the following advantages: the company has greater control over the use of the technology; it can more readily maximize its return on the investment over the long term; CDC may have access to new technologies if the partner is itself technologically sophisticated and thus be able to develop new product lines; it has access to more markets; it realizes economies of scale; it reduces its risk; and it conserves R and D and investment funds. As noted earlier, by joining forces with other firms, CDC has enhanced its competitive position. Its participation in Computer Peripherals and Magnetic Peripherals provides it with a strong position in a highly volatile area of the world computer market.

It is interesting to note the breadth of CDC's relationship with other firms. Its joint venture partners range from highly sophisticated computer manufacturers in the United States and Western Europe to firms in Rumania and Iran with little or no relevant experience. In the Rumanian case, it accepted a minority position as one of the conditions for achieving market entry. In the case of Computer Terminals of Iran, accepting minority ownership through a transfer of technology was preferable to committing funds in a political environment which leans toward nationalization. In short, CDC's approach to joint ventures has demonstrated a high degree of flexibility which has benefited the company without creating substantial problems to date.

Notes

1. Peripheral equipment includes such devices as tape transports, disk drives, line printers, card readers, data entry systems, terminals, and optical character readers. Peripheral equipment frequently accounts for up to 75 percent of the cost of a computer installation.

2. The central processing unit (cpu), or computer mainframe, is the portion of the computer that stores, retrieves, and processes data.

3. Software denotes computer programming instructions. By contrast, hardware refers to actual equipment, such as the mainframe or peripherals.

4. Raphael Tsu. "High Technology in China." *Scientific American,* December 1972, pp. 13-17.

5. "Virtual memory" allows a customer to use software programs as though a larger computer memory space existed than was actually available.

6. "Amdahl-Fujitsu Deal Now a Joint Venture." *Datamation,* June 1974, pp. 114-115.

7. See p. 19 of the firm's S-1 registration statement dated 1973, filed with the Securities and Exchange Commission under no. 2-48451.

8. See p. 33 of Amdahl Corporation's S-1 statement filed with the Securities and Exchange Commission, 15 September 1975, and registered as no. 2-54595.

9. "Amdahl-Fujitsu Deal Now a Joint Venture." *Datamation,* June 1974, pp. 114-115. Also "A Tyro Challenges IBM in Big Computers." *Newsweek,* 12 May 1975, p. 68.

10. Contained in Amdahl's S-1 statement to the SEC, 15 September 1975, and registered as 2-51595.

11. Ibid., p. 32.

12. "A Prophet Without Honor in His Own Country." *Forbes,* 15 November 1975, p. 24.

13. Contained in Amdahl's S-1 statement to the SEC, 15 September 1975, Exhibit 13(h), p. 3.

14. GE withdrew from the computer industry on relatively advantageous terms. It received $110 million in notes and $124 million in Honeywell stock, thus giving it 18.5 percent ownership of Honeywell Information Systems. When RCA and Xerox later abandoned their computer businesses, they incurred net losses of $210 million and $84.4 million, respectively.

15. Because of nontariff barriers, foreign firms were prevented from freely establishing computer manufacturing facilities in Japan. In light of this prohibition, Honeywell, GE, and other U.S. computer manufacturers entered into licensing arrangements with Japanese firms. The Honeywell-NEC agreement was signed in 1962. The GE-Toshiba license followed in 1964.

16. "The Gamma Invasion." *Fortune,* April 1959, p. 83.

17. Reportedly, de Gaulle still harbored resentment over Chrysler's takeover of Simca. See Thomas R. Branstein and Stanley H. Brown, "Machine Bull's Computer Crises." *Fortune,* July 1964, p. 155.

18. Gregory H. Wierzynski. "GE's $200 Million Ticket to France." *Fortune,* 1 June 1967, pp. 92ff.

19. "Staying the Course: Honeywell Vows It Will Stay in Computers." *Forbes,* 15 December 1975, p. 36.

20. Ibid.

21. "CDC in Iran Joint Venture." *Electronics News,* 15 December 1975, p. 1.

5

The Consumer Electronics Industry

Sector Overview

The U.S. consumer electronics industry has experienced a severe erosion of its world market position in the past decade. In large part, this decline can be attributed to aggressive and innovative Japanese competition. As a result, many multiproduct American firms in the industry have been forced to phase out certain product lines and sell off their technology. Foreign purchasers with capabilities for both high technical absorption and aggressive commercial expansion have been able to adapt required technology and in some cases to shift segments of production to lower-wage areas, thereby posing strong competition to other U.S. consumer electronics firms that retained a high domestic content.

Other active purchasing groups of U.S. electronics technology have been socialist economies and oil-rich developing countries. These two groups have been successful in negotiating highly effective management-service contracts aimed at developing internationally competitive enterprises. U.S. firms in the consumer electronics industry may face competition in traditional product lines in their export markets for components once these turnkey operations come on stream.

Japanese manufacturers have taken over an increasing segment of the growing U.S. market for consumer electronics products either directly through exports or by setting up production facilities in the U.S. or indirectly through manufacturing under license for U.S. firms. Matsushita Electric has become the world's leading manufacturer of electronic products and related household electric appliances. In 1963, imports constituted less than 7 percent of U.S. consumption of consumer electronics products; in 1974, they accounted for 42 percent of such consumption. During this period, exports of U.S.-produced consumer electronics products increased by 4.6 percent per annum, while imports rose by an average rate of 26.5 percent. From a 1966 peak of 110,000, employment of production workers declined by 14,000 in 1967 and declined an additional four percent every year thereafter. In the U.S. television industry, assembly plant employment dropped from a high of 42,920 in 1971 to 28,446 in 1975.

The Japanese consumer electronics industry matured behind nontariff barriers with the aid of substantial government subsidies. Japanese firms in this industry were exempted from antitrust legislation. Foreign investment was carefully restricted, and imports of consumer electronics products were subject to rigid quotas. As Japanese manufacturers grew more sophisti-

97

cated, they achieved substantial economies of scale in the production of specific items. Barriers to market entry at this point were lowered, but Japanese production costs were lower than those enjoyed by manufacturers in other countries. In addition to supportive government policies, Japanese success in consumer electronics exports can be attributed to sound management practices, high productivity of labor, imitation or adoption of product innovations, and shrewd marketing strategy.

When Japanese transistor radios first penetrated the U.S. market in 1955, a pattern was set that was often repeated in coming years. Though portable radios were originally developed in the U.S., the Japanese first mass-marketed them due to a gross miscalculation of their market potential by American manufacturers. Similarly, Japanese manufacturers penetrated the U.S. market with low-priced tape recorders and transistorized, portable monochrome TVs in the early 1960s, with solid-state high-fidelity systems and TV receivers in the mid-1960s, and with electronic hand-held calculators in the early 1970s. In each of these cases, U.S. companies continued to emphasize expensive and physically bulky "big ticket" products, such as home consoles in radios, TVs, and phonographs, in contrast with smaller, low-priced innovatively styled Japanese imports. The Japanese successfully mass-marketed these new products through such nontraditional distribution outlets as drugstores and discount houses. Because they identified U.S. market segments that were inadequately served by traditional consumer electronic products, the Japanese were able to stimulate substantial demand for their products.

In order to counter the cost-competitive aspects of the commercial challenge from abroad, U.S. manufacturers shifted much of their production to offshore facilities, primarily in the Far East. These plants were engaged in the production of labor-intensive components that were destined to be shipped for final assembly in North America. The Japanese, however, have also begun to rely heavily on offshore production, while Taiwan, Singapore, and South Korea have themselves become important suppliers to the U.S. market.

By 1974, all U.S. manufacturers had discontinued their domestic production of consumer electronics products, with the exception of color TVs. But even the exception proves the rule. In September 1977, Zenith announced a massive shift of component production to foreign facilities, which will cost 5600 Americans their jobs and will mean there are no longer any all-American television sets. Inescapably placed on the defensive, U.S. consumer electronics firms are concluding various management service and technology purchase agreements with foreign enterprises. U.S. TV makers, needing something to fall back on in the face of Japanese competition, have been acquired by larger, diversified concerns—Warwick by Sanyo, Philco by GTE Sylvania, Magnovox by North American Philips Corp., Admiral by Rockwell International Corp., and Motorola by Matsushita. As demonstrated by the latter case, the prevailing response of American TV manufacturers to Japanese competition is to join rather than fight the opposition.

Tremendous market potential exists in the emerging line of home videotape recorders (VTRs), but all manufacturing will be done in Japan. U.S. firms are reduced to marketing and retailing, under license, such products as Matsushita's Great Time Machine, its UHS System, and Sony's Betamax. The Japanese are careful not to release any crucial technology to American or other distributors and stand to receive substantial royalties as well as expanded manufacturing employment from what is expected to be a billion dollar VTR industry by 1980. The Japanese, who years ago began improving their technological skills with imitative and adaptive R and D efforts, have reached a stage where their innovative capacity is such that they no longer have to rely on technologies purchased from the U.S.

In large measure, the decline of the U.S. consumer electronics industry may be attributed to the failure of U.S. firms to innovate in such a way as to respond to rapidly changing market demands. As their technological advantage dwindles and their domestic labor costs rise, U.S. firms have begun to lose interest in manufacturing consumer electronics products at home. By the 1970s, the Japanese were clearly dominant both in terms of technological innovation and adaption to consumer preferences and in terms of planning and executing an aggressive marketing strategy.

Case Study: GTEI's Turnkey Plant in Algeria for the Manufacture of Consumer Electronics Products

Late in 1974, GTE International, Inc. (GTEI), a subsidiary of the General Telephone and Electronics Corporation (USA), signed a contract for the construction of a $232 million consumer electronics manufacturing plant with the Algerian enterprise SONELEC. SONELEC is the state corporation responsible for the development of an Algerian consumer electronics industry. The *produits en main*[1] contract requires GTEI to develop a vertically integrated operation that is virtually self-sufficient in component parts manufacturing (including semiconductors[2]), final product assembly, and product design-engineering.

The transaction is significant because of its size and its potential impact on the world consumer electronics industry. It is also illustrative of the scope of development goals Algeria has set for itself. Basically, SONELEC has purchased the capacity to become a significant factor in international consumer electronics. When the vertically integrated plant comes onstream, SONELEC will have a significant export potential in consumer electronics products. It will have an independent technological base that will be supplemented by related know-how during the life of the contract and allow the company to remain a state-of-the-art competitor. In addition, several hundred SONELEC employees are being trained in relevant technical areas so that the company will have a work force that can operate independently in the design and manufacturing areas. Finally, SONELEC will have a vertically integrated plant that will operate with limited dependence on foreign components and will provide existing Algerian companies in the same sector with a domestic source of previously imported component parts.

The GTEI plant is being built in Sidi-Bel Abbes (about 300 air miles southwest of Algiers) and will employ 5400 people. Once completed, it is expected to have the following production program on a one-shift basis:[3]

TV sets (black and white)	190,000
TV sets (color)	50,000
Radios (transistor)	410,000
Stereo/phono	50,000
Cassette players	60,000
UHF/VHF tuners	250,000
TV picture tubes (black and white)	238,000

The Sidi-Bel Abbes facility will be Algeria's second major consumer-electronics manufacturing facility. Another facility located at Blida (just outside Algiers) is privately owned and has a production capacity of 200,000 TVs and 100,000 radios annually. The plant employs 1500, and it is heavily dependent on foreign components. Over 50 percent of the sophisticated parts at present are imported. It has been speculated that production at the Blida facility will wind down as the SONELEC operation comes on-stream.

At the end of 1971, there were 160,000 TVs and 700,000 radios in use in Algeria according to one estimate. By 1974, it is believed that annual internal demand had reached 120,000 new TV units and 100,000 new radios per year. Thus, at present demand levels for this year, Algeria could meet its current needs in radios and would have a slight shortfall in TVs.

The new GTEI facility will be used to supplement the domestic shortfall in TVs and in consumer areas not previously served by the domestic industry, most notably tape and cassette recorders. In addition, it will supply most of the essential component parts for the factory and the other domestic plant near Blida. Much of the surplus production in finished products is probably intended for foreign markets.

Approximately $25 million of the contract price is earmarked for technical and managerial training in the U.S. Two hundred and sixty-five SONELEC employees will receive up to eighteen months classroom and on-the-job training in the U.S. and Europe. An additional 105 Algerian students will learn English and receive instruction in relevant technical undergraduate course work at New England colleges. SONELEC expects that this investment will provide the complementary manpower pool of executives and technical people it needs to maximize its capital investment in this new turnkey consumer electronics facility. Training for this industry's specific application instead of a generic educational program has been dictated by necessity and represents a departure from the more recent procedure in Algeria. Up until the present, there has been a major flaw in Algeria's state planning. It has limited its reliance on foreign technical and executive expertise at a time when it has had an inadequate domestic manpower pool. Consequently there have been operational difficuties, quality-control problems, and market development failures. At present, its industries are operating at only 30 to 60 percent of capacity.

From GTEI's point of view, the SONELEC contract offered a host of advantages and entailed acceptable levels of commercial and economic risk. First, the terms of the contract call for the transfer of technologies limited to the design and engineering of a specific manufacturing facility and production of specific products. The contract is executed over a finite period and tied to an existing state-of-the-art technology. Although the agreement requires GTEI to keep SONELEC informed about new technical developments, they will not necessarily be incorporated into the plant. Also, GTEI is making rapid advances in technologies relevant to the consumer electronics field and is incorporating these into its product lines and, therefore, will probably not suffer an erosion of its competitive position in world markets as a result of SONELEC production.

Second, the TV design provided by GTEI to SONELEC is not in compliance with U.S. standards.[4] This safeguard eliminates the possibility of exports to the U.S. market, without major product-design alterations.

Finally, for several years after the plant is on-stream, SONELEC will, under the terms of the contract, continue to be dependent on GTEI for some of the key ingredients needed in the component manufacturing and final assembly. Algeria is not a traditional export market for the U.S. electronics industry, and without this contract, GTEI would have virtually no export potential in this market.

In addition, GTEI was interested in winning this contract because it would serve as a sorely needed stimulant for its electronics division. The company, along with the rest of the U.S. industry, was faced with the prospect of idling installed capacity because of a dramatic drop in consumer demand and the increased competition from cheaper Japanese imports during the early 1970s. The contract assured continued plant utilization and retention of trained production workers for the company's operation in New England and in addition would provide business for numerous local subcontractors.[5]

In short, GTEI wanted the SONELEC contract because it would give them a substantial foothold in a market from which they otherwise would be totally excluded, and it would enable them to do so without jeopardizing their competitive position in world markets.

There clearly were substantial benefits to be derived from the SONELEC contract. And had GTEI not consummated the agreement, there were several foreign competitors who would have, including Thorn, Ltd. (United Kingdom), Nippon Electric (Japan), and IT and T Standard Electric (Germany). By not entering into the agreement, GTEI and other U.S. firms would have lost follow-on business for the supply of GTEI component parts (in addition to earnings from contingent goods and services) that are valued in eight figures per year over and above the basic contract. The company would also have forfeited future market opportunities in this sector of the Algerian economy. Once the plant is in operation, the Algerian government will close the border to competing imports.

The prospect that SONELEC will become a competitive factor in world markets anytime in the near future is slim. On the basis of Algeria's prior record in utilizing innovations in technology to upgrade products and the chronic delays in getting basic construction and design-engineering work implemented, industry observers predict that the new facility will not maintain technological parity for the time being. It is widely known that the Algerians have had considerable difficulties with other similar *produit en main* projects and that many plants are operating well below installed capacities and with the inevitable cost inefficiencies. The generally low level of skills found among Algerian industrial labor is a contributing factor, in that it inhibits the rate of absorption of acquired technology.

In addition, the added costs of upgrading (for retooling or re-equipping segments of the plant) would create a significant strain on SONELEC's projected project financing because of existing cost overruns. Finally, the Algerians feel the consumer product designs they are getting from GTEI will be good enough for the Algerian market from a functional standpoint for some time to come. If at some time in the future, SONELEC found it worthwhile to export, technological upgrading would become indispensable.

The GTEI-SONELEC project is a good example of the new modes of operation U.S. firms must enter into in order to maintain their position in overseas markets. Increasingly, they are forced to operate in difficult technical absorption environments—but significantly, the U.S. firm takes little or no financial risks in projects of this nature. For GTEI to refuse to meet SONELEC's terms would have meant not only loss of this contract opportunity but also loss of the Algerian market for electronic products as a whole. On the other hand, if the Algerians prove that they can make a commercial success out of a difficult technological transplant, GTEI may face future competition as a supplier of electronic components in Western Europe from a self-reliant Algerian industry.

Case Study: The Sale of Motorola's Color TV Manufacturing Assets to Matsushita

In May 1974, Motorola sold the vast majority of its color TV manufacturing assets and patents to Matsushita Electric Industrial Company, Limited. This sale largely ended the company's involvement in the home entertainment electronics industry; earlier it had phased out production of stereos, phonographs, tape players, and clock, table, and pocket radios. Like other U.S. manufacturers, the company's Consumer Electronics Division had had great difficulty retaining its competitive position in the face of rising imports from Japan, industrial overcapacity in the production of top-of-the-line units, and reduced U.S. consumer purchasing power. The most significant factor that affected Motorola's market position was Motorola's inability to match the competitive strength of the extremely competent domestic manufacturers, particularly Zenith Radio Corporation. Motorola could not match their research and development expenditures, distribution systems, and production facilities. Zenith's marketing strategies were also important elements. Finally, there was irrational pricing as well as overcapacity in the domestic industry.

In the early 1970s, the company's product lines comprised communications products (e.g., two-way radios and health monitoring equipment), semiconductors (e.g., transistors and integrated circuits), government electronics (e.g., aerospace communications and missile guidance systems), consumer electronics (e.g., color and black-and-white TVs and data displays), and automotive products (e.g., automobile radios and tape players).

Consumer electronics ranked third in terms of divisional contribution to corporate revenues, behind both communications products and semiconductors. In 1973, consumer products accounted for 17 percent of the company's $1.4 billion in sales and resulted in a loss after taxes of approximately $3.8 million. In the preceding four years, the sale of TVs was responsible for up to $20 million of pretax losses.[6]

In contrast, Matsushita had over $4 billion in sales, 30 percent of which derived from its color-TV product line. The company had previously marketed a broad range of consumer electronic products in the U.S. under the Panasonic brand name, and it continues to be the world's largest manufacturer of consumer electronics. Exports accounted for 18 percent of its total sales in 1973.

In Japan, Matsushita was known as a "follower company." That is, it was not known to be a frequent innovator of new products; rather it was thought to have achieved a strong market position by simplifying the design and production of items already proven in the market. These improvements usually facilitated either price reduction or product differentiation. In each case, the company was able to attract an enormous clientele through its highly successful promotional efforts.

In 1973, Motorola had less than 7 percent of the U.S. color TV market, compared to Zenith Radio's 24 percent, RCA's 22 percent, Magnavox's 7 percent, and Sears' 7 percent (Sears was the brand label under which two manufacturers' TV products were sold). Considering that Motorola had earlier been a leading innovator in the color TV industry, it was surprising that its market share was so low. In 1966, the company was the first to introduce solid-state color TVs. This technological improvement was primarily a substitution of transistors for vacuum tubes and was believed to result in a more dependable and more durable product. Marketed under the name Quasar, sales growth of this machine greatly exceeded the industry's average—40 percent in 1972, for example, compared to 20 percent for the industry as a whole.

Because of lack of plant capacity and a general unwillingness to invest large additional sums of money in the production of color TVs, Motorola chose to employ solid state technology only in its 25-inch console, on which the company had concentrated much of its sales effort. At a time when the market for portable color TVs was rapidly expanding, in at least partial response to Japanese firms' U.S. advertising efforts, Motorola proceeded to emphasize these $600 models. Although the unit volume of these sales continued to show an upward trend after other manufacturers had begun to mass-produce portable solid-state color TVs in 1972, Motorola's profitability on its color-TV product line seriously eroded. The company, therefore, made a costly shift in emphasis away from console models. Extending the Quasar technology into different product segments was expensive, as was the incremental advertising effort it necessitated.

A more critical consideration, however, in terms of market position was Motorola's inability to keep pace with the solid state technology. It did not make the necessary research and development expenditures in solid state technology. Motorola sold the solid state idea to the public, but it was overtaken by its competitors, Zenith and RCA, when it failed to make investments in advanced solid-state designs. As a result of its decision, Motorola found itself in the position of not being able to catch up with Zenith and RCA.

Many factors bore on the company's decision to sell its color TV operations to Matsushita. The U.S. color TV market was rapidly becoming saturated. With only a 7 percent share of this market, Motorola could anticipate low-level replacement sales. Another factor in the decision was the belief that the industry might shortly require large additional R and D outlays in order to stimulate further sales. At this time, Motorola was coming under pressure from other portions of its product line. The decline in U.S. automobile sales was hurting the company's automotive products division. With the war in Vietnam near an end, the sale of government electronics was on the decline. Semiconductors were profitable but highly volatile business. U.S. consumer purchasing power was also diminished by the prevailing combination of recession and inflation.

Motorola and Matsushita agreed that the purchase price would be slightly above book value, which was determined by an independent audit to be approximately $100 million. The purchase included Motorola's TV production facilities at Franklin Park, Pontiac, and Quincy, Illinois, although Motorola would continue to operate the Quincy Plant until 1976. Motorola's leased assembly operations in Canada and its relevant product inventories were assumed by Matsushita. The U.S. company retained its plant in Taiwan.[7]

Matsushita was authorized to utilize the Quasar brand name. In fact, Motorola's former color TV operations were reorganized into a subsidiary of Matsushita which become known as Quasar, Inc. Although Matsushita has probably benefited by the use of this internationally recognized brand name, the company has maintained its Panasonic-brand products and its previous channels of distribution. Of its consumer electronics product lines, Motorola has retained only data displays.

After the public announcement of the intended sale, both Zenith Radio and Magnavox filed complaints with the Federal Trade Commission. These firms argued that the sale to Matsushita would give that firm sufficient market power to hamper free competition in the U.S. consumer electronics industry. The U.S. government ordered a delay, during which time Motorola was asked to search for a buyer for its TV operations. None of the U.S. firms that might have been interested were able to make the purchase, however, because of possible antitrust violations. Thus, the sale to Matsushita was allowed to occur. As part of the agreement, the Japanese firm agreed to retain the work force at the Illinois plants for an unspecified period of time.

Had the sale been made prior to 1970, a strong outflow of unique technology would have occurred. Between 1966 and 1970, Motorola's Quasar technology was without equal. The company did not, however, maintain its technological leadership in the 1970s. It is therefore probable that at the time of the change of ownership Motorola had little unique technology to transfer to Matsushita's color TV operations. Whatever technological advantages Motorola may have enjoyed in color TV production were no doubt transferred to Japan shortly after the acquisition. Representatives of other firms in the industry have suggested that there was a small initial outflow of unique technology followed by a more significant inflow of technology from the Japanese firm.[8]

One aspect of the transfer of technology was the granting of certain patent rights by the U.S. firm. Without these patents, the production activities in the Illinois plants could not have continued as before. The acquisition of these patents did not necessarily provide Matsushita with a technological advantage over its U.S. competitors, however.

Matsushita's purchase of Motorola showed that it was serious about obtaining a larger share of the U.S. market. The Japanese market had become saturated to some extent, with three-fourths of Japanese households owning color sets. Matsushita executives evidently hoped that the new Motorola subsidiary, through Quasar's prestige, would provide a strong entrée into the console market.

Moreover, the Japanese conglomerate, as the world's largest consumer electronics firm, certainly was in a better financial position than Motorola to make substantial investments in technological improvements. Such innovations as three-dimensional pictures, flat screens, miniaturized sets, and home video-tape recorders are perhaps necessary to spur lagging industry sales.

Circumstances that led to the Motorola-Matsushita deal were indicative of the state of the U.S. consumer electronics industry in the late 1960s and 1970s. Prior to the dramatic rise of Japanese imports, the industry had served American consumers with traditional products. Japanese firms successfully identified market segments that could be better served with innovative products. Once these opportunities were identified, U.S. manufacturers were slow to react. Their unresponsiveness led to a rapid deterioration of the market position of U.S. firms and a significant increase in concentration within the industry.

The actual transfer of technology probably kept jobs in the U.S. (though Matsushita has phased out the Pontiac, Illinois plant); U.S. jobs may have been lost if Motorola had abandoned its color TV operations without having found a buyer. Japanese technology is reportedly now flowing into the Illinois plants, strengthening the American production base and providing benefits to American consumers. Motorola, having divested itself of an unprofitable product line, is presumably a healthier company because it has reoriented its business into areas in which it is better able to compete. Still, the Motorola case offers little consolation to those who had hoped American firms could vigorously respond to the Japanese import challenge.

Case Study: RCA and Corning Glass Works' Transfer of Color Cathode-ray Tube Technology to Poland

In May 1976, RCA Corporation and Corning Glass Works separately completed negotiations with UNITRA, the Polish state-owned electronics corporation for the sale of technology and equipment for the manufacture of color cathode-ray tubes. Although the two companies are providing equipment and know-how that is complementary and essential for the plant, the companies have not formed a joint venture or coordinated their sales efforts and activities in any way. Negotiations were initiated at different times, separate contracts were signed, and separate financing was arranged.

The combined value of the contracts is at least $124 million. Both companies expect to earn additional income through the sale of parts and additional services during the life of the contract. In addition, the contracts call for the payment of royalties on tubes produced and exported. The plant is expected to be completed in the late 1970s. It will be located at Piaseczno, a suburb of Warsaw. Initial production output capacity will be approximately 300,000 units per year of 21-inch diagonal picture tubes. There is a production output potential of 600,000 units annually. Total TV production (monochrome and color) in Eastern Europe and the USSR is approximately 7 million units per year. The USSR will continue to be the exclusive supplier of color units until the Polish facilty is on-stream. The new agreements will reduce Polish dependence on the Soviet Union and provide an alternative source of tube supply for other Eastern European countries.

Both Corning and RCA have had extensive dealings with UNITRA and other East European and USSR state enterprises. Prior transactions involved the sale of related technologies and parts and equipment. Both companies contend that a sale of technology is made to these countries only when it is not possible to make unit sales. Nevertheless, while the possibility of selling completed cathode-ray tubes did not exist, both firms believed that they would have significant market potential in terms of the sale of component parts over the ten-year life of the contract. The components, such as aperture controls and support pins, although they are not technically sophisticated, are produced by complex machinery. These parts are more economically procured by UNITRA on a unit basis than manufactured because of the need to invest in additional plants and equipment. In addition, Poland had alternative sources for these technologies.

Interestingly, RCA has the technology and capability to manufacture a complete color tube system without outside assistance. The company possesses the technology, engineering know-how, and manufacturing experience. It does not, however, have the specialized experience and know-how that Corning has acquired in the manufacture of glass specialty products. UNITRA's decision to contract with two different companies for the complete system was more expensive but was apparently based on a desire to obtain the preeminent technology for each of the major component parts of the tube system: the electron gun and the glass envelope and picture screen.

RCA held the initial talks with UNITRA over three years ago. Negotiations were of a general exploratory nature until the spring of 1975. At this point, negotiations became more focused. RCA was eliminated as the supplier of the whole system, and talks were limited to their role in the production of the nonglass bulb components. Although Corning had been in general contact with UNITRA for a period of three years, exchanges were limited to inquiries on capabilities or the availability of state-of-the-art technologies. One year prior to the signing of the contract and at about the same time RCA was eliminated as a supplier of the total system, Poland entered into detailed discussions with Corning. Hard bargaining commenced three months prior to the contract signing.

Corning's contract was for the sale of a manufacturing package for production of the glass envelope component of the picture tube. The sale covered equipment, assistance in purchasing foreign manufactured products that were not covered in Ex-Im Bank financing, engineering assistance, and training in the U.S. and at the factory site in Poland. Technical training is limited to teaching requisite engineering skills for the construction and operation of the facility.

The contract price for the sale by Corning is $55,521,900. This figure does not include an additional fee for factory warranty check-outs, identified as start-up costs. (This amount will be an additional $4 million, which will be apportioned between RCA and Corning. It is not subject to Ex-Im Bank financing.) Sixty percent of the major contract price is for equipment and 40 percent is for software. The contract extends over a period of ten years and includes a provision for regular updating of technology related to the design of the glass envelope sold to Poland. The agreement does not include different or new designs and related technology.

Financing was essential in this sale. Corning was not willing to assume any long-term accounts receivable. Although the financial arrangements in terms of mechanics were the same for both Corning and RCA, they were separate transactions involving separate agreements and negotiations. Financing was arranged by UNITRA through Bank Handlowy w. Warszawie S.A., an agency of the Polish People's Republic. Two sources of funding were utilized: a private U.S. banking consortium led by First National City Bank (FNCB) and the Ex-Im Bank. Financing or loan guarantees by the Ex-Im Bank are limited to materials of U.S. origin in the project.

Corning received a 10 percent cash down payment of approximately $5.5 milion, which was not financed. The FNCB consortium provided a loan for 35 percent of the contract, or $19.43 million. Interest on this loan is at a floating rate of up to 1 5/8 percent above the prime rate on the London Interbank rate plus set fees not to exceed 1 3/4 percent. The Ex-Im Bank financed 55 percent of the project, or $30.5 million, at a rate of 8 3/4 percent. The repayment will be in fourteen semiannual installments beginning in October 1979. The first five installments and part of the sixth will be applied to the repayment of the private financing. A portion of the sixth and the last eight installments will be used to repay the Ex-Im Bank loan beginning in April 1982.

As noted, Corning will earn additional income over the life of the contract by the sale of individual component parts and on royalty payments on each unit produced for export outside Eastern Europe. The company has been granted the unusual right to audit the books of the UNITRA plant when it becomes fully operational. This annual audit will provide a means of checking the accuracy of the royalty payments, which are tied to production output.

The company has provided a warranty that lasts until the last Corning technical employee has finished his work at the plant site. This time is expected to be within three months of the factory going on-stream. Further, the warranty is effective for a trial operation period. This cost is separate. During the trial period, raw materials utilized in the bulb manufacturing will be imported from Corning-controlled sources. This provision was made because the equipment is designed and built with specific tolerances and for raw materials of a particular standard. The company will not accept responsibility for operations utilizing raw materials over which it has no quality control.

Corning made the sale of the equipment and technology for several reasons. First, it was unable to penetrate the Polish market on a unit sales basis. Second, foreign competition, particularly in Japan, would have made the sale if Corning had not. In fact, Corning was under considerable competitive pressure from the Japanese, who were offering a system with much more attractive financing terms than were privately and publicly available in the U.S. Third, the company believed the sale was an opportunity to tie this market to U.S. technology.

UNITRA's choice of Corning was based on its having the preeminent technology, an established record in the glass envelope industry as an innovator and as a company with experience in transferring technology. In addition, the Poles anticipated substantial language difficulties with Japanese-speaking technicians. Apparently, neither the Poles nor the Japanese have a substantial base of employees with reciprocal language proficiencies. Corning, as well as RCA, has available large pools of technical labor with relevant language capabilities, and Poland has a comparable work force fluent in English.

Corning has chosen in at least one recent case to make an equity investment in a technology transfer project with an East European country.[9] No consideration was given to this type of arrangement in the case of Poland because the company does not have a market share that can absorb the additional capacity for the tubes to be produced. This constraint is partly the result of an established and stable market position upon which it would be hard to improve. But, in addition, mechanical features of the Polish tube make it incompatible with some of the largest, established Corning markets.

Corning believes it will not be adversely affected in the U.S. or overseas by this sale to Poland. The technology provided is not state of the art but is basically the same technology utilized in the U.S. It provides Poland with a new and significant technology base, one that would require a substantial investment of time and money to develop independently. Of course, this assertion presumes that no alternative technology sources existed. In addition, Poland has been successful in developing on its own a fairly sophisticated black-and-white tube manufacturing capability. Therefore, the transfer cannot be construed as providing a unique capability.

Corning believes it has limited an immediate Polish threat to its markets. It has done this in several ways. First, the tube design does not meet U.S. and some European x-ray emission standards. Consequently the finished product cannot be exported to these markets. Second, the technology provided will not permit the unit price of the finished product to be competitive with U.S. costs. Third, the company has developed unique new designs and technology that will further undercut a Polish cathode-ray tube's market in the U.S., even in the face of corrections in x-ray emissions and costs. Corning recently introduced a new glass bulb that is lighter, cheaper, and easier to produce than the one used currently in the U.S. Of course, the current U.S. production model already has a significant competitive advantage over the Polish tube. The new unit has a 100-degree angle and pyramid shape that allows for a 2.5 inch decrease in the depth of the tube. The standard tube design utilizes more space and provides a 90-degree angle. It is less energy-efficient than the new design. The company expects this model to go into production in the U.S. within the next year. In the interim, it is in the process of making further refinements in this technology and expects to remain a major innovator and leader in glass tube technology for electronics applications.

RCA sold the Poles the equipment, technology, design, and know-how to produce the electronic gun component of the cathode-ray color tube. The reasons for the sale were similar to Corning's. RCA could not successfully market its tubes in Poland; RCA was aware of discussions for the purchase of this technology from competitors in Japan and Western Europe, most of whom were former licensees; little or no adverse impact on RCA's worldwide markets was expected and an immediate, positive impact on employment for the company and its subcontractors was foreseen. Finally, the company maintains that it will retain its position as a technological innovator and leader in this field and be in a position to maintain its market position and advantage.

The sale, which provides for engineering and technical training, is limited to the know-how needed to construct and operate the factory. But the Poles may be expected to use this new technology and their apparent technological sophistication as a basis for new innovations and to seek new markets outside of Eastern Europe. Whether they actually will do so remains open to speculation. RCA, however, does not expect great market potential for these tubes because of the high demand in Poland's domestic market, which will probably absorb the total output for five to seven years. No apparent consideration has been given to Poland's need to earn foreign exchange, in Western markets. While the contract contains a royalty agreement based on unit output, it is at a flat rate and untied to various export markets.

The contract covers a ten-year period and, like Corning's, provides an update on technology during this period but only for this particular style of electron gun. A warranty agreement covers replacement parts and a start-up period. RCA expects to make limited sales of unit parts over the first part of the contract. Within the ten-year period, however, it expects that UNITRA will begin producing its own parts and manufacturing equipment. In other words, it will have a duplicative capability.

UNITRA's decision to buy the RCA system was based on many of the same factors mentioned in the Corning discussion. Additional points included a desire to deal with contractors from one country, the compatibility of RCA and Corning technology and parts and the actual use of these components together over many years in the U.S., and RCA's reputation as a technology leader and its considerable experience in the sale and transfer of technology to other countries.

These sales will have an immediate and medium-term positive impact on the U.S. economy in the following ways: the Corning Glass Works's contract will generate 1841 person-years of labor for U.S. workers, in addition to 837 person-months of management services; and the RCA component of the project will result in 2228 person-years of work. The figures for both companies do not take into account additional fees and royalties for technical services, the warranty check-out, equipment procurement, and the use of patents and know-how. On the other hand, the potential negative impact on the U.S. economy and employment as a result of the technology sale can be expected to be minimal.

Though both Corning and RCA apparently believed that UNITRA is unlikely to export tubes and/or TV sets to Western markets (because of internal demand, product design, etc.), this may not be the case. The need for foreign-exchange earnings to service Poland's substantial debt (over $9 billion) is sufficient incentive for UNITRA to become innovative and to redesign these tubes for an export market. As demonstrated in other projects negotiated by the Polish government with U.S. firms (e.g., the proposed GM-Polmot agreement), their interest is to enter into production and comarketing agreements under which export of a percentage of the output can be used to pay for the technology acquisition. Because of continuing balance-of-payments deficits, there is a good likelihood that U.S. firms will continue to experience pressures to assist the Polish economy to expand its export earnings. In the past, Polish authorities have not hesitated to sacrifice domestic consumption for the sales of foreign-exchange earnings. Moreover, experience in other industries, such as that of U.S. machine tool manufacturers, indicates that the Poles have not hesitated to exploit a new technology resource even when the design is proprietary and protected by patent.

Notes

1. *Produit en main* means literally "product in hand." The term refers to a "turnkey-plus" contract that includes training of all levels of plant personnel through the top management level and aims ultimately at a self-sustaining capability to manage and operate a production facility without further assistance from outside (foreign) sources.

2. GTEI tried to convince SONELEC officials that manufacture of semiconductor devices for the volume of demand in the Algerian market would not be economical, but the Algerians want self-sufficiency and either are willing to pay the cost premium for below international-scale production or hope to expand production through exports in the foreseeable future.

3. Output has been planned on a one-shift basis. The production figures cited could be increased by up to 25 percent by the addition of a second shift. Production volumes would have to be increased to lower the costs to internationally competitive levels, but the Algerians have indicated that they are not interested in export markets—at least for the time being.

4. The Algerians will be using the French TV system based upon different broadcasting frequencies and picture line densities.

5. During this period, the Boston metropolitan area, the location of the consumer electronics facility of GTEI's parent company, GTE, was faced with the highest level of unemployment since the Great Depression of the 1930s.

6. "Dead End." *Forbes,* 1 March 1974, p. 27.

7. At this point, Motorola has independently decided to withdraw from the black-and-white television business and to wind down its Taiwan operation by liquidating its inventory and equipment. Toward that end, Motorola continued to operate its plant in Taiwan for approximately one year after the sale of its color TV business to Matsushita. The facilities were eventually sold to a U.S. manufacturer of components for consumer electronics products.

8. This comment is based on remarks made at the National Academy of Engineering's "Seminar on Technology Transfer as a Result of Foreign Direct Investment in the United States," panel discussion on *Electronics, Computers, and Scientific Instruments,* New York, 3 February 1976.

9. Corning initiated such a transaction in Hungary. This is a joint venture for the manufacture of blood-gas analysis equipment and related instrumentation. In this case, Corning made a cash investment and supplied the key technology in exchange for 49 percent of the company stock. Dividends are guaranteed in dollars.

6 The Chemical Engineering Industry

Sector Overview

It is necessary to preface this overview with a brief description of what we mean by "the chemical engineering industry" because it does not have a definition that is as widely recognized as, say, the aircraft industry. There are essentially two kinds of firms in the industry: the chemical process firm and the engineering-construction firm of chemical, petrochemical, and refining complexes. We recognize that there can be some overlap in these two categories and that some firms engage in both activities (but frequently at an arm's length from one another). Interestingly, firms develop a reputation for one kind of ability; companies such as Lummus, Stone and Webster, Kellogg, and UOP are better known for their abilities with process technology, and firms such as Fluor, Bechtel, McKee, and Parsons are better known as project managers.

An important distinguishing characteristic of the firms in this industry is that they generally are not end-users of the processes they develop or the design, engineering, and construction services they provide. Their business is to commercialize research and development results into processes and engineering services from technologically based proprietary positions in various fields. The policy they follow in managing their technological assets is significantly different from that of a product firm. The latter can earn a return on those assets only after it has developed a product and successfully commercialized that product in the marketplace. The product firm, therefore, will necessarily be protective of its technological resources and capabilities. The chemical engineering firm, on the other hand, earns a return only when it commercializes its technology and, therefore, views its business as that of selling newly developed technology as extensively as it can and reinvesting a portion of its profits in developing new generations of technology. There is little concern for the competitive potential of technology purchasers.

The case studies prepared on chemical engineering firms demonstrate the industry's drive to develop and earn returns on its technology in the design engineering, process design, and construction dimensions of industrial systems. Commodities produced by these systems rarely pose a competitive threat. Scenarios include the transfer of project management techniques by the Fluor Corporation, the export of technology for

115

catalytic converters for automobile exhausts by UOP to several Japanese firms, and the sale of proprietary process technology for acrylonitrile production by Sohio to the PRC. As is further discussed in this sector overview and in considerable detail in each case study, the nature of the technology and the timing and mode of its transfer are conditioned, in large part, by the corporate considerations and the purchasing environment.

The Role of the Chemical Engineering Firm

Firms that provide a major portion of the engineering and construction services required for a major complex generally have a long and close relationship with the client. There is usually one project manager who supervises the whole operation from basic engineering design to eventual production. He is responsible for coordinating local subcontractors, ensuring that the project complies with local building codes, and seeing that the plant is completed on time and within the established budget. After the engineering design is established, the contractor takes care of detailed engineering, canvasses vendors of equipment and material, purchases technically satisfactory equipment, transports material to the job site, and erects the plant in accordance with the design package and the customer's special requirements.

The client will often not take charge of the operation until the turnkey process is complete, although it will work in close coordination with the contractor throughout the engineering, purchasing, and construction phases. Thus, when production begins, the client's local management will be able to manage the chemical plant autonomously but often will remain dependent on the contractor for specialized manpower and equipment purchasing services long after production has begun.

At the present time, there are some twenty to thirty major firms that are capable of taking full responsibility for the design and construction of chemical complexes throughout the world. They are supported by approximately a hundred other firms that handle only parts of a major complex.

The chemical process firms in the industry constitute one of the most dynamic and technologically intensive industrial groupings in the country. The processes they develop are used to produce a related series of items, including chemicals of all kinds, paints, pharmaceuticals, synthetic fiber and rubber, detergents, fertilizers, plastics, and photographic supplies. Their R and D represents one of the largest privately financed R and D efforts in the world. U.S. government funding has been limited to defense and related areas—such as synthetic rubber development during World War II.

The chemical process firm has more limited contact with the client than the engineering-construction firm. This usually comes in the form of negotiations for patent rights to a particular process technology. Firms with an edge in process technology often do much of the original development work themselves but may also work closely with chemical and oil companies who have developed processes. Then, acting partially as agent for the process developer, these firms may adopt the process to their client's requirements.

The growth of chemical engineering-construction firms received its main impetus during World War II. At that time, there was strong world demand for synthetic materials, both petroleum and nonpetroleum based. The European and Japanese chemical industries were virtually destroyed during the war and during the early postwar years the U.S. supplied 90 percent of chemical products and process technology. This state of affairs soon changed when U.S. policy makers decided to give incentives to U.S. industry to rebuild West European and Japanese industry. The chemical engineering-construction company fulfilled the needs of war-devastated economies that required implantation of a chemical industry.

The Industry's Management of Technological Assets

Successful management by chemical engineering firms of their technological assets is a function of at least three variables: the form in which the technology is sold, the competitiveness of the particular technology in world markets, and the rate at which a technological asset erodes. As regards the first variable, a chemical engineering firm argues that while it provides a client detailed instructions, drawings, specifications, blueprints, demonstration, and so on, of how a particular technology works, it withholds the logic of the technology—the essential ingredient in developing an innovative or even duplicative capability of a particular technology. The industry contends that although it is possible for some of the recipients of its technology to determine the core of the process, it would be uneconomical in terms of time and money for them to attempt it.

Maintaining competitiveness in international markets requires the constant upgrading of technology by plowing back retained earnings into R and D. Fierce competition exists among process and engineering firms in the industry and each one must maintain its technology at the state-of-the-art level if it is to remain competitive.[1] When a firm's profits drop, however, one of the first areas whose resources management finds easiest to cut is the R and D division.

The third variable that determines the degree to which technological assets are successfully managed—the rate at which the asset erodes—in many ways relates to the first two variables. Erosion can occur when the recipient of a particular technological process eventually masters its ingredients and comes to understand why those ingredients behave as they do or when other firms develop a more competitive technology.

The speed with which a recipient absorbs technology and develops a duplicative or innovative capability in a particular technological process depends on the level of its own technical and scientific resources. For example, one firm that we interviewed—UOP—estimated that after the technology has been licensed, it has an eight-year lead over a Brazilian client but only a two-year lead over a Japanese client. Western European nations and the People's Republic of China are particularly adept at duplicating engineering technology. This capability was fully demonstrated when UOP first visited China in the early 1970s to discuss future business transactions. The process team was shown an exact duplicate of its own—albeit, an eight-year-old model—catalytic cracker and platforming refinery of a 3000-barrel-a-day capacity apparently copied from one that UOP had previously sold to Cuba.

Those within the chemical engineering industry are quite confident that their respective firms are capable of maintaining a sufficiently safe lead over their clients. At the same time, however, many of the newly industrializing countries and socialist nations aspire to close that gap. The development of indigenous design, basis and detailed engineering, and construction capabilities is an essential element of a strong capital goods industry, which, in turn, is basic to an industrialized economy.

Firms in the industry that provide design, engineering, and construction services labor under similar pressures. Clients—particularly in developing areas—are equally anxious to develop their own design, engineering and construction abilities and to free themselves of dependence on foreign firms for this know-how. The degree to which they can realize these objectives is a function of how the company operates and, again, the absorptive capabilities of the client. At this point in its development, Algeria, for example, is dependent on the foreign firm for all major services: design and engineering, construction, procurement, and start-up. The Soviet Union, on the other hand, has achieved a fairly high stage in the development of its capital goods industry and its technical manpower and, therefore, is mainly interested in design and engineering services and some technical assistance during start-up. It insists on handling most of its own procurement and construction and does all the civil (foundation and utilities) and structural (building frames) engineering. Some feel, however, that this insistence on doing their own civil and structural engineering is based more on national pride and a determination to achieve self-sufficiency than on an accurate assessment of their capabilities in this area.

The U.S. Industry's Competitive Position
in World Markets

A survey of the last twenty years of activity in chemical engineering showed that in 1956, U.S. firms held near complete dominance of the field—engineering some 91 percent of the 221 oil, petrochemical, and agricultural projects then being engineered and constructed. In mid-1976, there were 2266 projects underway, and U.S. firms, though clearly the most important national group, were handling only 47 percent of them. As stated earlier, the origins of many of the foreign competitors lay in U.S. reconstruction efforts and industry assistance to Japan and Western Europe following World War II. A detailed breakdown of these activities and the main foreign nations participating in them is shown in Tables 6-1 and 6-2.

The ten-fold growth in the number of chemical engineering projects underway worldwide has taken place mainly outside the U.S. The number of projects in the U.S. grew by a factor greater than four during the last two decades, whereas in other countries the growth factor is seventeen. While U.S. firms continue to do more of the non-U.S. projects than firms from any one other nation, the foreign firms are now clearly providing increasingly tough competition for the U.S. firms offshore. The leading U.S. chemical engineering firms have established subsidiaries in various parts of the world in order to be able to bid more competitively on certain design and construction projects in host countries and in third countries. (In the U.S. itself, U.S. firms still have almost complete dominance; in 1976 they handled over 90 percent of the projects underway.)

The number of petrochemical projects has increased much faster than the number of refinery and agricultural chemical projects. There are now twenty-three times more petrochemical projects underway than there were in 1956, whereas there are seven times as many refinery projects and nine times as many agricultural projects. U.S. engineering firms do more projects in all three of these fields than firms from any one other country, but it is in the petrochemical projects that the U.S. firms have lost most to foreign competition. Whereas U.S. firms did over 90 percent of the projects in all three fields in 1956, they now do only 40 percent of the petrochemical projects but 47 percent of the agricultural chemical projects and as much as 53 percent of the refinery projects.

Despite the fact that U.S. chemical engineering firms no longer completely dominate the world scene, they are still clearly the single most important national group in it. U.S. firms handle five times as many projects as the firms of any other single nation—and five times as many projects now as they handled in 1956.

Outlook for the Future

Analysis of the construction boxscore data and the interviews revealed a growing concern that foreign competition is intensifying. Changes in relative international prices (due in part to different rates of inflation and in exchange rate adjustments) affect the competitive situation—currently for the better for U.S. firms whose unit costs are now below those of Europe

Table 6-1
Analysis of Projects Handled by Chemical Engineering Firms Worldwide

No. of Projects Underway in		Handled by Firms from
1976	*1956*	
1. All projects		
1070	202	U.S.
231	4	Italy
209	0	Japan
194	3	Germany
132	1	France
95	10	United Kingdom
335	1	Other countries[a]
2266	221	All countries
2. Refinery and gas-handling projects		
558	143	U.S.
141	4	Italy
96	0	Japan
45	0	France
23	1	Germany
22	5	United Kingdom
176	1	Other countries
1061	154	All countries
3. Petrochemical projects		
378	33	U.S.
136	1	Germany
93	0	Japan
71	0	Italy
69	5	United Kingdom
61	0	France
126	0	Other countries
934	39	All countries
4. Agricultural chemical projects		
104	26	U.S.
35	1	Germany
26	1	France
20	0	Japan
19	0	Italy
4	0	United Kingdom
33	0	Other countries
241	28	All countries
5. Projects underway in North America[b]		
562	123	U.S.
8	5	United Kingdom
6	0	France
1	0	Germany
1	0	Italy
0	0	Japan
41	2	Other countries
619	130	All countries

Table 6-1 (cont.)

No. of Projects Underway in		Handled by Firms from
1976	1956	

6. Projects underway in Western Europe and Japan

119	37	U.S.
97	0	Japan
95	4	Italy
70	1	Germany
42	5	United Kingdom
32	1	France
80	1	Other countries
535	49	All countries

7. Projects underway in resource-rich countries[c]

161	25	U.S.
76	0	Italy
71	0	Japan
20	2	Germany
10	0	France
10	1	United Kingdom
30	2	Other countries
378	30	All countries

8. Projects underway in socialist countries[d]

79	0	U.S.
63	0	France
40	0	Germany
35	0	Italy
28	0	Japan
22	0	United Kingdom
20	0	Other countries
287	0	All countries

9. Projects underway in "other countries"

109	13	U.S.
64	0	Germany
35	0	Italy
22	0	France
20	0	Japan
12	0	United Kingdom
183	0	Other countries
445	13	All countries

Note: These analyses and tables were prepared by Developing World Industry and Technology. The source of the data is Gulf Publishing Company's "Construction Box-Score," published in its journal *Hydrocarbon Processing* and its predecessor *Petroleum Refiner.*

Note: The construction boxscore in Tables 6-1 and 6-2 lists all the projects that have been publicly announced and shows which engineering firms are active in them. In using the boxscore data, consideration was given only to those projects that were being engineered by engineering firms—as opposed to those being engineered by the company owning the project. Approximately 5 percent of all projects reported as being engineered in the 1976 boxscore were being handled by the staff of the project's owner. The definition of a

Table 6-1 (cont.)

"project" follows the definitions in the boxscore; thus each of the major individual units of a refinery or a petrochemical complex is listed as a project.

The major deficiencies of these analyses are that the breadth and accuracy of reporting has clearly increased in the twenty years being examined and it is likely that some foreign activity in 1956 has not been picked up and that all projects are counted equally although they probably differ widely in value, amount of engineering involved, and uniqueness of the technology employed. Nevertheless, the analysis does provide a good snapshot of the activity of chemical engineering firms in different fields and areas.

All subsidiaries of engineering firms are counted with the country of the parent firm.

This explanation also refers to the data in Table 6-2.

aFirms from other countries were mainly from Holland, Spain, Brazil, Mexico, India, Australia, and South Africa.

bNorth America here means the U.S. and Canada.

cThe eleven members of OPEC, Brazil, and South Africa.

dUSSR, Eastern Europe, and the PRC.

and Japan where the currencies have been revalued. This situation can reverse again with another round of exchange and price adjustments.

At the same time, the world oil and chemical industry is poised for a major round of new building, and although more of the pie might go to foreigners, there may be a much larger pie to slice. There is even concern among some U.S. firms that they will not be able to expand fast enough to meet the potential demand for their services. If the U.S. should decide on a national energy policy, a quantum jump in local building would be necessary.

Nevertheless, the firms interviewed are worried about the continuing threat to their once unquestioned position of dominance. There is scant possibility of meeting the threat as in the past—by setting up wholly owned subsidiaries abroad. Not only is there a growing resistance to the wholly owned subsidiary abroad, but there have been a number of recent examples of subsidiaries breaking away from the parent and entering into direct competition with it.

When they do go out to meet the foreign competition head on, U.S. firms often feel at a disadvantage, without the government-provided support that their competitors seem to enjoy. They are not able to offer the same financial sweeteners, they lack detailed commercial and political intelligence, and they find themselves also competing with other U.S. firms. On the other hand, the foreigners usually do not compete among themselves on international bids.[2]

Another major concern for many companies is the political and legal developments that will affect the need for their services in the U.S. itself. The lack of a comprehensive energy policy, and the on-again-off-again nature of environmental controls are particular worries. As these companies rightfully contend, to effect any one of many policy options now being con-

Table 6-2
Analysis of Growth of Chemical Engineering Projects Worldwide

| | Underway in 1976 | | Underway in 1956 | | Growth[a] |
	Number	% of Total	Number	% of Total	1956-1976
1. By type of project					
Refinery and gas handling	1091	48	154	69	7
Petrochemical	934	41	40	18	23
Agricultural chemical	241	11	28	13	9
All projects	2266	100	222	100	10
2. By location of plant site					
North America	619	27	130	59	4
W. Europe and Japan	537	24	49	21	11
Resource-rich areas	378	17	30	14	13
Socialist regions	287	13	0	0	0
Other regions	445	20	13	6	34
All areas	2266	100	222	100	10

Note: For further explanation, see the discussion at the bottom of Table 6-1.
[a]Number of times.

sidered at all levels of government will take massive investments of money, manpower, and time. They do not feel inclined to make these investments unless they are assured that the policies dictating them will be followed through.

Conclusions

Despite the reservations and concerns of chemical engineering firms as to their flexibility in carrying out international operations and their competitive position in world markets, the industry still is a dominant force on the world scene. Demand for its services is growing. The Fluor Corporation foresees a $25 billion business in the design, equipping, and construction of chemical and oil process plants by the 1980s.

A major problem arises, however, when one considers that much of the demand for the services of chemical engineering firms today comes from resource-rich countries with severely limited local markets. The strategy of these countries is to increase earnings by processing the raw material at the source of its extraction. In the past, refining and petrochemical plants were built to meet demand for their output and were located at the source of that demand. Today, in an increasing number of cases, the plants are being constructed on the basis of supply. This proliferation of manufacturing and processing capabilities, especially in the Middle East, Brazil, and to a smaller extent in Korea and Taiwan, can create worldwide marketing problems. Once the new plants are on-stream, their commodities will, by necessity, compete in world markets because they cannot be consumed by local demand. Countries that have served as export markets for U.S.-processed petrochemicals, synthetic fibers and rubber, and so on, in the near future will be net exporters themselves of these same commodities. When one adds to this scenario the fact that the new producers have immediate access to the raw materials necessary for the manufacture of a broad spectrum of commodities at much cheaper prices, one can conclude that the competitiveness of U.S.-based production of these commodities may be severely eroded.

Case Study: UOP, Inc.—a Product, Process, and Engineering/construction Firm

UOP, Inc., is a highly diversified firm that develops and sells petroleum and petrochemical processes, products, and services; construction and engineering services; and a variety of engineered products. Most of its seventeen divisions earn a return on investment only through the sale of products; one earns returns through the sale or licensing of technology; and another one produces income through design, engineering, construction, and related services. Each of these three kinds of operations—product, process, and engineering-construction—manages its technological assets in a different way so as to maximize its returns and competitiveness in world markets. In addition to briefly describing the scope of UOP's operations, this case study will delineate the attitudes and policies toward technology by focusing on three of the firm's divisions: the Automotive Products Division; Process (R and D and licensing arm); and Procon (design, engineering, and construction subsidiary).

The company's basic business is to commercialize research and development results into processes, products, and services from technologically based proprietary positions in the fields of energy, environmental improvement, and high-value-added manufactured products. The processes include a wide range of oil refining and petrochemical processes for the manufacture of fuels and petrochemicals such as para-xylene, a key chemical in the manufacture of polyester materials. UOP's products include—to name but a few—copper tubing, catalysts, copper-clad and unclad laminated plastics, suspension and fixed seating for trucks and other vehicles, catalytic converters for automobiles, seats and galleys for commercial aircraft, specialty chemicals, air pollution control equipment, pipe motion compensators, and membrane separation systems for water and waste effluent purification. It also provides a full range of design, engineering, and construction services for petroleum refineries, petrochemical and chemical plants, and gas-processing facilities, as well as assistance in plant personnel training, start-up, and operation.

UOP places strong emphasis on research and development in all areas of its business. In 1974, UOP was issued 896 U.S. and foreign patents. Currently, UOP has about 10,000 active patents worldwide.

In 1975, 50.5 percent of UOP was acquired by the Signal Companies, a $2 billion conglomerate based in California. Signal is also the owner of Mack Trucks and the Garrett Corporation. It is too early yet to tell what impact this will have on UOP, but it is generally thought that Signal was attracted to UOP's technology and R and D capabilities.

UOP Process Division and Procon, Inc.

The UOP Process Division, which is the descendant of the original UOP, develops and licenses petroleum refining and petrochemical processes. It designs refineries and petrochemical plants and provides services such as start-up, operation, and inspection. The UOP Process Division also manufactures catalysts, additives, and inhibitors. At present, the Process Division employs about 1600 people worldwide with 1461, or 91 percent, located in the U.S. The Process Division licenses more than thirty-five different patented processes. Approximately 200 new units using UOP licensed processes have been designed each year since 1969; some 3000 units are now operating in over eighty countries today. About two-thirds of the UOP Process Division revenues today derive from its refining technology and one-third from its petrochemical technology.

One of the most outstanding achievements in UOP's history was in petroleum refining when, in 1947, the research laboratory developed a new catalytic reforming process using a catalyst containing platinum. Platforming, as the process was named, allowed for the reforming of low-grade naphtha into high-octane motor fuel. Introduced commercially in 1949, there are now over 500 such units installed throughout the world. UOP Process Division designs and catalysts have been constantly upgraded and expanded through its R and D efforts with the aim of maximizing yields of high-octane gasoline, aromatics, and desulfurized fuels and reducing operating costs.

The primary services of UOP's construction subsidiary, Procon, Inc., include mechanical design, engineering, procurement (including inspection, expediting, and shipping), and construction. Related services include technical and economic feasibility studies, technology evaluations, licensing arrangements, environmental services, assistance in arranging international financing, project management, engineering and management consulting, engineering and construction management and supervision, plant personnel training, start-up and operation, and contract maintenance.

Despite optimistic indicators, in terms of contracts completed and backlog orders, Procon has had problems registering profits in recent years. Several explanations were offered by the firm for Procon's financial situation. First, and perhaps most important, Procon is paying dearly for contracts negotiated at a fixed cost several years ago that are now reaching the completion stage. That its contract negotiators are now taking a tougher stance on this issue is illustrated by the fact that at the end of 1974, some 72 percent (in dollar value) of the work to do was in cost-plus contracts, while only 13 percent was cost-plus at the end of 1973. Secondly, competition is more intense than ever in the process engineering and construction field.

The inability or unwillingness of the U.S. government to arrive at national policies and programs in the areas of energy and the environment has also affected significantly the level of business of Procon as well as the UOP Process Division. These units are somewhat dependent on legislation and national policy to create an effective climate for new capital expenditures by their customers in the U.S. New construction of petroleum refineries in the U.S., for example, will probably be deferred until Congress or the executive branch or both set forth and implement a new energy policy.

Finally, the worldwide recession induced by the energy crisis severely inhibited and, in some places, stopped altogether development of new projects in the mid-1970s. Procon pointed out, however, that new opportunities have been created as a result of the necessary world energy realignments.

It is unlikely that the UOP Process Division and Procon will ever be combined as a working unit even though they more naturally complement each other's functions than any of the firm's other divisions. Synergism between the two is not desirable for several reasons. For example, it is generally felt that more clients will seek their respective services if not obligated to use both the technology and the construction services of the same firm. In many cases, in fact, the client may not even want construction firms to know they are in the market for a construction job, a fact that would certainly be signaled by the knowledge that it was negotiating for a license to a particular technology. Maintaining the independence of the two divisions is significant in that most other large chemical engineering firms do combine the two services.

Most of UOP's petroleum research in new technologies is applied and not exploratory. Its R and D procedure is generally as follows. UOP marketing specialists identify which processes need upgrading and conduct market surveys on the extent and profitability of this need. If the market survey indicates a substantial need for the improved or new product or process, the technical and scientific staff of the Corporate Research Center begins work on its development. The research is then transferred to the Process Division, where it is further developed by engineers and designers into a pilot project.

According to UOP Process Division management, its licensing arrangements attempt to incorporate the following elements and qualities. First, UOP Process avoids setting up an arrangement that would cause it to compete with the licensee. Second, it protects the licensee against infringement of the patent rights of others. UOP also gives a complete guarantee to the licensee that the process will function properly. Finally, it agrees to make available to the recipient all improvements and innovations it has made in the technology. The recipient's improvements and innovations in the same technology are made available to UOP's other licensees of the same technology.

The only way in which UOP Process can earn a return on its investment is by commercializing technology. In managing its technological assets, it is of little concern whether the client is a domestic or foreign firm or whether the client seeks the technology to acquire simply an operative capability or a duplicative or innovative capability. The process technology firm seeks only to earn returns of which a portion can be plowed back into development of successive generations of technology. This policy permits it to remain competitive among other technology suppliers as well ahead of its clients.

Although Procon—the engineering-construction division of UOP—conducts no R and D in technology, it has developed considerable expertise and know-how in project management systems and engineering, which to a greater or lesser extent are transferred to a client each time the division undertakes a job. Developing an indigenous capability in these skills is a high priority in many newly industrializing countries seeking to expand national capital-goods industries and in most socialist economies. Procon is under constant pressure to improve and upgrade its cost efficiency, scheduling systems, advanced engineering skills, overall project management, and so on, to maintain competitiveness and world market share. It is in this way that it manages its technological assets.

UOP Automotive Products Division

The product firm earns a return only when it develops a product that successfully competes in the product market. Unlike the process or engineering-construction firm, then, it is protective of the technological know-how required to produce the product. This section will briefly trace the transfer of UOP catalyst technology from its traditional petroleum market to an entirely new area, automotive exhaust pollution control. In a technical sense, the two markets are similar. The scientific principle of catalysis is identical and the method of manufacturing is basically the same. From the marketing point of view, the two areas could not be more dissimilar. The market for UOP's petroleum products has been created through the logical process of product improvements and time-proven efficiency. On the other hand, the market for automotive catalysts was created in a public arena by politicians caught between an unstoppable force—environmentalists—and an immovable object—the auto makers' desire to move slowly and to utilize proven technology.

The basic components of gasoline are hydrogen and carbon. The fuel is combined with air (i.e., oxygen and nitrogen) and injected into the combustion chamber, where it is burned. Ideally the products of combustion should be heat, carbon dioxide, and water vapor. However, an automobile engine is not perfect, and it produces harmful pollutants: unburned fuel or hydrocarbons (HC), carbon monoxide (CO), and oxides of nitrogen (NO_x).

In 1952, it was discovered that HC and NO_x were the primary constituents of Los Angeles smog. Subsequent research confirmed the deleterious effects of the CO emitted from autos. As the size of the automobile fleet in the U.S. and overseas grew, so did the pollution problem, and as the problem grew, so did the demand for a solution. As the environmentalists grew in power, government responded to their pressure for legislation to solve the problem. Auto makers and other private companies started searching for a solution.

Even in the 1950s, the potential market for automotive pollution controls was substantial. New car sales were 7.2 million in 1955 and there were 44 million cars on the road that year. With this stimulus, industry explored the areas they knew best. The auto makers worked on engine operating parameters—spark timing, exhaust gas recirculation, fuel delivery, engine design, and many others. Carburetor manufacturers studied modifications in their product. UOP and others in the 1950s took the different and more radical approach of catalytic control.

A catalytic converter for auto exhausts looks very much like a muffler. The exhaust gases go into one end of the mufflerlike can, pass over the catalyst contained in the can, and go out the other end. As the gases pass over the catalyst, unburned hydrocarbons (i.e., unburned gasoline) and carbon monoxide react with oxygen, which is also present in the exhaust gases. These reactions convert the hydrocarbons and carbon monoxide into harmless carbon dioxide and water vapor. The purpose of the catalyst is to speed up the reactions to make sure that they are completed in the split second that the gases are in contact with the catalyst.

The active part of the catalyst is a noble metal such as platinum or palladium, which is spread very thinly on a ceramic base. This ceramic is inactive in the reaction and serves only to carry the noble metal and to extend the surface of the noble metal in contact with the gases. Two forms of ceramics are used: small pellets and the so-called monolithic form. The monolithic ceramic catalyst support is a single piece of honeycombed ceramic made by extruding the soft ceramic through a die before baking it. It is important that in each form the ceramic has a rough and very porous surface over which the noble metal can be spread.

In 1964, UOP had developed a pelleted catalyst and converter (catalyst container) that was both durable and effective. The device was approved by the state of California for installation on 1966 and future model-year cars sold in the state. The Los Angeles smog problem was solved, or so it appeared.

At this point, idealism and advanced technology met head on with the real world. The environmentalists had wanted someone, anyone, to solve the problem. The politicians had responded by outlawing smog and directing government administrators to force industry to develop a solution at no direct, or at least readily traceable, expense to the consumer. The catalyst industry developed one solution, but the technology was ahead of its time. The auto makers would not accept it because it was too new and unproven for mass production and widespread consumer use. Besides, Detroit had developed its own mechanical solution. The politicians did not want to utilize catalyst technology, even though it may have been best, because they would have had to force each and every motoring, taxpaying voter in the state to pay the cost of installation. California officials opted for the hidden cost of the auto makers' solution.

UOP and others, including auto makers, continued catalyst R and D but with a much lower priority. However, the pollution problem did not decrease. It grew and became international in scope. Needless to say, the environmental movement kept pace.

On the last day of December 1970, the now famous Clean Air Act Amendments became law. Among the numerous provisions was a 90 percent reduction of automotive HC, CO, and NO_x from the 1970 level by 1975. Within weeks, the major auto makers and potential suppliers initiated high-priority research and development programs. By 1972, the catalyst appeared to be the most practical solution to the auto makers' dilemma. In 1973 Chrysler, Ford, and General Motors contracted with various manufacturers to begin construction of plants to supply automotive emission control catalysts. Though the auto makers fought the standards, they had to be prepared with catalyst suppliers so that they could meet them, if and when imposed. Overseas auto companies with U.S. markets soon followed Detroit's lead. Japan passed a strict domestic emission-control law in 1972. After nearly twenty years of work, a new version of an old business was created.

UOP signed supply agreements with Chrysler, Nissan (Datsun), Toyota, and Fiat. Subsequently, Toyo Kogyo (Mazda), Daihatsu Kogyo, and Porsche agreed to purchase catalysts from UOP. The reasons behind these contracts vary from customer to customer, but there are some common elements. UOP was recognized as a leader in the petroleum catalyst field, and its record in automotive emission control was well established. The company had existing or proposed production capacity. They had the necessary personnel and facilities to enter into an intensive research and development program to perfect their existing catalyst products to meet customer specifications. In the case of Fiat and Porsche, UOP had the added capability of being able to design the catalyst container. As in any business decision, however, the most important factor was the price-value relationship. Those who believed UOP had the best combination of capabilities bought their product, while the others went elsewhere.

The UOP Automotive Products Division has a substantial share of the auto emission control catalyst market, and they expect to increase their penetration. This anticipation is based on three beliefs. First, their potential automated-catalyst manufacturing process gives UOP a price-quality advantage for the several products they make. Second, they have designed and patented monolithic and spherical catalytic containers, which are produced by subsuppliers or under license. Third, UOP's research and development program keeps them abreast of or slightly ahead of the competition.

Unlike the UOP Process Division, the Automotive Products Division's main business is selling products, not licensing technology. For each unit of catalyst or each converter and catalyst sold, a specific profit factor returns to the company. Because of the patents and the proprietary nature of the products, their existing technological advantage is not depleted by each object sold in the U.S. or overseas.

Case Study: Sohio's Sale of Acrylonitrile Technology to the People's Republic of China

Throughout its twenty-five-year history, the People's Republic of China has exhibited wide swings in its receptivity to foreign technology. The year 1970 marked a return to a more permissive technology import policy after the violent upheavals and intense antiforeign campaign during the Cultural Revolution of the late 1960s. During the earlier periods of relative openness, the Chinese selectively purchased European and Japanese plants and equipment, primarily as prototypes for learning and copying. Since 1970, however, they no longer confine purchases to prototypes and have bought large numbers of complete plants and industrial complexes to increase output in a half-dozen basic industries, primarily metallurgy, petrochemicals, and energy.

The Standard Oil Company of Ohio (Sohio)—the smallest of four companies using the Standard Oil name in the U.S.—was one of the first American firms to sell technology to the PRC during the current period. The four Standard Oil companies were among the thirty-four companies that resulted from the breakup of the Standard Oil trust some sixty years ago, and they are now strong competitors. Sohio is ranked about 110 in *Fortune*'s 500 list. For a company of its size, it is unusual that Sohio does no manufacturing overseas. It has been eminently successful in its research, particularly in the chemical field and has chosen to earn a financial return on its intellectual property outside the U.S. by licensing rather than manufacturing. Within the U.S., Sohio has also followed a policy of licensing its technology to all qualified parties, including competitors. This open license policy has resulted in larger royalties, which in 1974 totaled around $40 million.

The process Sohio licensed to the Chinese was for the production of a chemical called acrylonitrile from propylene, ammonia, and air. The process has proved to be Sohio's major licensing success. It is recognized to be the most economical route to acrylonitrile. The chemical is used to produce acrylic fibers sold in the U.S. under such familiar trademarks as Orlon and Acrilan. Large amounts of acrylonitrile are also used in the manufacture of a plastic called ABS (acrylonitrile, butadiene-styrene resin). It is an exceptionally strong material and can be used in appliances and automotive parts.

The process Sohio licensed to China was to be used in a plant to produce 50,000 metric tons of acrylonitrile annually. In a typical licensing of the process, Sohio provides the basic technical information, operation manuals, operator training, and start-up assistance. The rest of the engineering and construction services are provided by an engineering firm. In the PRC arrangement, Sohio turned to Niigata Engineering Company, Ltd., of Tokyo, which had built ten Sohio process acrylonitrile plants. Asahi Chemical Industry Co., Ltd., a licensee of Sohio's acrylonitrile process, was also brought into the group to provide start-up assistance, operator training, and other services that Sohio would normally provide but that U.S.-Chinese regulations might prohibit. Asahi was also to act as general contractor for the project and to be responsible for general supervision, financing, and other matters.

The Sohio contract was negotiated in ten days in Peking by two Sohio executives, but Chinese negotiations with the Japanese for the plant itself continued for seven months. Sohio's request to the Office of Export Administration (OEA) of the Commerce Department to export the technology was submitted in June 1971 and approved in late 1972. This was the first U.S. license for China that the OEA had approved since the late 1940s. Both contracts were concluded with the China National Technical Import Corporation in early March 1973, and the plant sale was approved by the Japanese government on 2 May. Sohio was not told the purpose for which the acrylonitrile monomer would be used, nor did the Chinese specify the plant location (it is known now to be located near Shanghai).

The terms of payment for the acrylonitrile technology—the value of which is about $8 million—are noteworthy. Early in the project it was decided that the license agreement should be between Sohio and Asahi, as the general contractor, with a sublicense agreement between Asahi and the China Technical Import Corporation. Payments to Sohio would come directly from Asahi, and were not conditioned upon receipt of payments from the Chinese. Thus it was Asahi who guaranteed Sohio's payments.

At the start, Sohio requested a lump sum payment based on the design capacity of the plant and the Chinese readily agreed. The Chinese traditionally have an aversion to paying high interest rates and wished to establish a payment schedule minimizing interest charges. This payment schedule was rather easily resolved. Royalty payments for use of Sohio's technology would be paid in U.S. dollars through the Japanese.

Concomitant with the license agreement, a contract to supply Sohio's Catalyst 41 for the new plant was also signed. This recently developed catalyst gives China one of the world's most modern waste-reducing techniques, which increases production 40 to 50 percent over that of earlier-generation catalysts.

Aside from the license fees, the Chinese will pay the equivalent of $29 million for the Japanese plant. The sales terms call for payments in semi-annual installments over five years and involve Japanese Export-Import Bank credits at 6 percent—typical terms in other recent plant sales from Japan to China. The contract also includes acetonitrile, cyanic acid, and waste-water disposal units based on Asahi Chemical technology, which were expected to be on-stream early in 1976.

China's importation of the complete acrylonitrile plant represents just one of its three favored forms of acquiring foreign technology. The other two—industrial fairs or technological exhibitions and procurement of one- and two-of-a-kind prototypes—have not proved as efficient or complete as vehicles of technology transfer, although they are still quite popular.

A few technology exhibitions were held in China in the 1960s. They were stopped by the antiforeign campaign of the Cultural Revolution. Between 1971 and 1975, however, some thirty-eight were sponsored by both socialist and nonsocialist countries. Largely in response to Chinese urgings, the nature of the fairs is more educational than commercial, and the exhibitors tend to be the most prestigious within their respective industries. The technologies featured are usually the latest and most sophisticated the exhibitors have developed, and in some cases, prior to the fair they had never been demonstrated. The exhibits typically feature hundreds of technical seminars and industrial firms, backed up by demonstrations and displays; great quantities of technical data and glossy catalogs, translated into Chinese, are freely available.

In short, the exhibitions greatly facilitate China's search for relevant foreign technology, and perhaps more important, they provide Chinese technicians with detailed instruction and demonstration by highly qualified, top-level executives at virtually no cost. The extent to which the Chinese are able to absorb the lessons and how astute they are in eventually purchasing the technology are difficult to assess. Only a few things can be learned from looking at a display model, hearing it described and reading a technical sales brochure. A more rewarding route is to purchase the model for the purpose of analysis and reverse engineering. Such models can be bought very cheaply at industrial fairs because the exhibitors feel that they may thereby obtain an official entrée into the Chinese market. In addition, the return shipping cost is high enough to justify the cheap sale.

Prototype copying has limitations, too—although there are numerous recorded examples of successful Chinese prototype copying, without external assistance (the Massay-Ferguson 35 hp tractor is one of the earliest examples, and the Hasselblad 500 C/M camera one of the most recent). A major and difficult requirement of prototype copying is that the model must be duplicated at the right level of sophistication—with respect to design engineering, fabricating skill, machining precision, and materials application—relative to the level of competence attained by the Chinese themselves. If the model is too advanced, reverse engineering becomes difficult. Lacking the original design and manufacturing data, the copier must recreate the basic blueprints, the detailed engineering or wings, and most important, the materials specifications. More ambitious efforts to achieve a series production run, rather than simply to fashion a single duplicate, magnify these problems and create a multitude of new ones. The acquisition of efficient, high-volume production methods and advanced management technology is indispensable. Furthermore, large-scale production requires design standardization to achieve perfect interchangeability of parts and components and expertise in production tooling, plant layout, materials and work scheduling, and quality control.

The copier's task is formidable and extremely time-consuming, and in the end, he has merely demonstrated that, with patient effort, he can slavishly copy an existing design. The prototype reveals only what was produced—not why. An understanding of the logic or rationale behind the original designer's choices is prerequisite to improving or adapting the prototype. The Chinese, to date, have not been nearly as successful as the Japanese at finding the right combination of copying the designs of others and originating their own, and the proper evolutionary sequence of moving from one to the other.

In large part, official recognition of the problems inherent in acquiring technology via trade fairs and prototype copying has led the Chinese to import complete plants in recent years. This form of technology import, complete with technical data and advisory assistance, has served to ease gradually China's shortcomings in design and management technology. The major thrust of the new plant acquisitions centers on a few basic areas that are necessary for the country's growth and for feeding and clothing its people: the industrial fundamentals of steel, power, and petroleum and petrochemical industries—(chemical fertilizers for agriculture, man-made fibers for the textile industry, and petroleum-based plastics for numerous purposes).

All the plants to be delivered employ the most modern high-technology production equipment available, and it is expected that they will enable China to "leap forward" in production efficiency and product quality in the few industries they affect. Indeed, one industry that can be expected to take great strides in the near future as a result of China's importation of complete plants is man-made fibers. The Sohio-Asahi acrylonitrile facility is one of several such plants for which the PRC has signed contracts with private firms since 1973. A technical mission visited Japan in March-April 1972 to inspect vinylon and polyester fiber plants, and the following year saw orders placed in quick succession for man-made fiber equipment. A contract was signed in March 1973 for the purchase from Kuraray of a plant with an annual capacity of 33,000 tons of polyvinyl alcohol (the raw material for vinylon fiber) and 66,000 tons of vinyl acetate using Kuraray and Bayer processes.

In May 1973, Peking ordered from Toray Industries and Mitsui Shipbuilding a plant to produce annually 25,000 tons of polyester polymer by a Toray technique based on an improved DCD process. Also, in that same month, a plant to produce 90,000 tons of vinyl acetate and methanol a year was purchased from a consortium composed of Speichem, Humphreys and Glasgow, Rhone-Poulenc, BASF and L'Air Riquide. In July, Mitsubishi Petrochemical and some other Japanese firms sold China a 60,000-ton high-pressure, low-density polyethylene plant that will use an improved version of a BASF process. A similar plant, with 180,000-ton capacity, was ordered from Sumitoma Chemical and Ishikawajima-Harima Heavy Industry in September. A petrochemical complex ordered from France in September 1973 included an 87,000-ton polyester plant to be supplied by Rhone-Poulenc Textiles. A 35,000-ton polypropylene plant was ordered from Snam Progretti of Italy (using Standard Oil of Indiana technology) later in the same year. In February 1974, Peking signed a contract with Teijin for the supply by 1976 of a plant to produce 40 tons a day of polyester staple and 8 tons of filament. About the same time the Chinese completed negotiations for a polyester fiber plant of similar capacity from Toray. Around November 1974, Kuraray was awarded a contract by the National Technical

Import Corporation for a 45,000-ton polyvinyl alcohol plant to be set up at Chungking. Several catalyzer plants are also being built to complement the existing manufacturing facilities in the various stages of fiber production.

A watershed occurred in the PRC's textile industry around 1965, at which time the country's oil production substantially increased. Petrochemical-based fiber production was greatly enlarged while a shortage of timber caused the cellulosic fiber industry to be relatively neglected. (China's cellulosic fiber plants seem to be mostly small-scale local plants employing low-grade technology and equipment.) By that time, as well, China had demonstrated that it had mastered the techniques not only of man-made fiber manufacture, but of man-made spinning, weaving, printing, dyeing, and other processes. These developments were partly indigenous and partly borrowed from Japan, Great Britain, and other countries.

In reviewing any set of figures on the PRC, one must keep in mind that they are usually estimates and that they frequently vary markedly from another source's estimates. Nevertheless, several indicators allow one to safely state that China's man-made fiber industry has rapidly expanded. Industry journals indicate that China's production capacity for man-made fibers in 1976 increased by as much as 400 percent over the actual production figure for 1974.

Domestic crude oil production will assure China of the feedstock needed for its man-made fiber and other expanding petrochemical industries. By 1973, having achieved a production level of 1 million barrels a day, China was self-sufficient in oil production and was exporting 7 million barrels to Japan. By year-end 1974, it was estimated that China's oil production rate had increased to 1.4 million barrels a day, and its exports to Japan that year were 28 million barrels. Japanese sources estimated that China's 1975 oil exports to Japan would be 55 to 70 million barrels, depending on its construction schedule and port-facility expansion project.

Assessments by Western oil experts estimate the PRC's recoverable oil reserves at 25 billion barrels (compared to 34 billion barrels in the U.S.). Some Japanese put the total possible Chinese reserves at 70 billion barrels. According to a 1975 Joint Economic Committee Report, by 1980 China will be producing 3.8 million barrels a day of crude oil, which should permit continued growth of its oil exports to earn badly needed foreign exchange.

In addition to controlling its own source of oil, the PRC, through the purchase of several complete man-made fiber plants such as the Sohio acrylonitrile facility, will have gathered sufficient technology within a few more years to build its own synthetic fiber plants on a large scale. Many in the industry have no doubt that this prediction will come true, given the impressive Chinese record of absorbing and duplicating Western technology and engineering.

136

China's population alone could consume much of this expanded production of synthetic fiber, and there has been a slight trend in this direction. At the same time, however, the government could select this sector as a major source of foreign exchange through exports. China is currently the second-largest exporter of cotton textiles in the world. Unless China prices its goods substantially below world market levels, however, they will not be competitive due to extremely high U.S. tariffs imposed on synthetic textile imports from non-MFN (most favored nation) countries. An increasing amount of synthetic fabric from China is being sold to Hong Kong and Macao where it is processed into clothes and then exported to the U.S. under MFN tariff status. The Australian textile industry, suffering from cheap imports, has recently insisted that the government seek voluntary restraints on Chinese exports. The PRC, to date, has refused to take action along these lines.

It is unlikely that China's production of synthetic fiber will compete directly in the U.S. market against domestic producers in the near future. A more likely scenario is that PRC textiles will compete with U.S. exports in Far Eastern markets. Foreign exchange earned through exports is necessary to pay for the expensive technology China has imported in recent years, including Sohio's acrylonitrile facility, and textiles may provide an appropriate source.

Case Study: The Fluor Corporation's Transfer of Project Management Techniques to Saudi Arabia

The Fluor Corporation, one of the largest and most successful engineering companies in the world, is headquartered in Los Angeles and has worldwide operations. Its main line of business has been the chemical and petroleum processing industries, particularly in the design and construction of plants totally under its responsibility. Total responsibility for handling a project has become a hallmark of Fluor and is the theme of this case study. It can be said that Fluor's concepts of total-responsibility project management and the techniques it uses have become a technology that to some extent is transferred on every job it undertakes.

On the average, Fluor does about one-half its business abroad. A dramatic recent example of this trend is that in 1974, Fluor obtained its first billion-dollar contract—in the U.S. as part of the trans-Alaska pipeline—and in 1975, it obtained its second billion-dollar contract, this time outside the U.S.—for the SASOL II plant to extract oil from coal in South Africa. Its biggest contract, signed in June 1975, was for a $4 billion gas-collecting and processing plant in Saudi Arabia.

Fluor transfers technology of all types abroad through its foreign operations. Not only does it transfer hundreds of engineering technologies every time it builds a plant for a customer, but it also involves foreigners in its project management systems and so transfers the technologies it has developed in that regard as well. Fluor's foreign subsidiaries are staffed and managed largely by local personnel who are trained in Fluor methods and who often contribute to the development of new and improved methods. (A detailed description will be given later in this report of these technologies and how they are transferred.)

The SASOL II and Saudi Arabian gas projects will provide Fluor with an opportunity to test its advanced project management techniques in locations that are remote from its headquarters and in which Fluor has not had recent large-scale operations. The testing will be valuable to Fluor because, if successful, it will put Fluor ahead of its competitors in proven ability to handle large projects in remote areas. Improving its techniques by using them is a policy that Fluor prefers to exercise on all of its contracts.

Fluor does not license its project management techniques as such. It delivers the techniques only as part of the total-responsibility package of services that it prefers to supply. Through its Bonner and Moore subsidiary, it does offer for sale computer software and consulting services that might form part of the project management techniques. It would be difficult, if not impossible, for a purchaser of these discrete parts of the total technique to reproduce the total technique. In this way, Fluor protects its methods; and in any case, through their constant use and improvement, Fluor is able to make them rapidly obsolete.

Fluor prefers to maintain complete control of its foreign subsidiaries and always assumes the dominant control on its foreign construction work. In this way, it is able to control access to its management technologies and to those items of proprietary engineering technology that it has developed.

However, in some cases, such as that of its Iranian subsidiary (and until recently its Taiwanese subsidiary), Fluor has found that the only way to establish a local operation is through partnership with local interests. One of the reasons that these subsidiaries appear to do less well than its 100 percent owned subsidiaries may well be that they do not have the same freedom of access to Fluor's latest techniques.

Fluor is keenly aware that its engineering and construction work can also give it access to technologies that may be new to it. While this is not a major consideration in Fluor's bidding on contracts abroad, such access is regarded as an advantage. In describing the SASOL II contract to stockholders, Fluor chairman, J. Robert Fluor, pointed out that the company will have extensive exposure to the most advanced technology in the world (developed by South Africa) for extracting coal reserves. Fluor's plant will incorporate that technology on a scale far larger than has ever been attempted and will greatly enlarge the company's technological expertise. The chairman further indicated that the technology can be made available in the United States.

Fluor's efficiency in project management is built around the task force approach in which design, engineering, procurement, and construction groups report to a single person—the project manager. He is totally responsible to Fluor management and to the client for all aspects of the job, including cost control and scheduling, as well as the business and technical details.

The task forces are staffed to meet the needs of each job. The staff of the task force will change in strength and composition as the job moves through its various phases, but a core team of project engineers and the project manager will stay with it until completion.

Once the client has selected an engineering firm, activities begin in earnest on the three main phases of the work, namely, the design and engineering, the procurement of the equipment and other supplies (cement, steel, etc.), and the construction work itself.

The three phases run simultaneously with one another, but each will reach its peak activity at a different time. Design and engineering will peak about 20 percent of the way through the total job, with procurement peaking near the 40 percent mark and construction at the 80 percent point. But even though design and engineering are the main concerns at the beginning of the job, construction will have commenced in a modest way with such operations as site clearing. Likewise, at the end of the job, when construction is the main focus, there will be residual design and engineering work as last-minute modifications have to be accommodated. Procurement work starts at the beginning of the job with the initial canvassing of potential suppliers of equipment and other items to determine their interest in the job, their current capacity position, and their likely prices, all of which enter into the process of supplier selection. Procurement moves to a shipping and delivery phase after the suppliers have been chosen and the orders placed, and it is still in evidence in the last phases of the job to correct deficiencies in the supplies and to acquire spare parts and other warehouse items for the client.

Fluor's two key management tools for project planning and control are designated with acronyms that are descriptive of the techniques: FACT (Fluor Analytical Cost Trending) and FAST (Fluor Analytical Scheduling Technique). Both techniques are computerized. FACT is of greatest importance at the beginning of a project, as decisions made then have the most effect on the project's final cost. Early cost estimates are constantly updated as new data are received and the cost trend of each item is plotted before firm commitments are made. Any change affecting cost can be quickly evaluated and reported. A daily check of cost trend versus budget is kept so that the project manager can take corrective action in good time. FACT provides the project manager and client with computer-printed reports of accumulated labor costs, material commitments and expenditures, and a summary report of project costs and forecasts.

FAST, on the other hand, uses a computer-assisted critical path diagram to provide a graphic plan of project progress. It is initiated by a step-by-step identification of each activity that is to be performed to complete the project and the sequence and time required for each activity. These are arranged mathematically on the logical "best" plan, which then becomes the norm for evaluating actual performance. If necessary, the basic plan can be modified. The computer-printed reports show the project manager and client the effect of each actual event and activity on the plan as each occurs so that early corrective action can be taken when necessary. An important feature of FAST is that it enables the project manager and client to analyze the effect of a proposed technical or procedural change on the project schedule before the change is implemented.

Fluor also has a scheduling tool called PROMPT (Progress Reporting of Material Procurement and Transportation), which supplements the overall FAST schedule by providing specific details on purchased equipment and materials. At new job sites, Fluor installs a computer terminal to provide rapid access to this program, as well as to FACT and FAST, and thereby to allow the management on-site (both Fluor's and the client's) to be intimately involved in the control of the project.

The construction manager will have the best of both worlds if he can successfully blend the skills and experience of his company with those of the local people. He will often try to maximize the involvement of the local people, especially the subcontractors, who will thus become trained in his firm's ways and will be available for future work. A fine line has to be observed in training the local contractors between making them efficient subcontractors in the future and preparing them to be potential competitors. Generally, in the design and construction of sophisticated modern chemical and petroleum plants, there is not much danger of the local subcontractor becoming a competitor, except in highly developed areas, such as Western Europe. For this reason, Fluor will usually hire all its construction workers directly for work there and avoid the use of subcontractors entirely.

An example of how Fluor is using these techniques in foreign locations—and so to one degree or another is transferring the technology abroad—is its very successful work for the National Iranian Oil Company (NIOC). Fluor built an 85,000-barrel-per-day oil refinery near Tehran in 1968 for NIOC. Almost before it was completed, the Iranians saw that demand for oil products was increasing more rapidly than they had expected and decided to build a similar refinery adjacent to the first plant.[3] To save time, the same engineering-construction team of Fluor and the German firm

Thyssen Rheinstahl Technik was employed. Many of the project, process, design, purchasing, and mechanical equipment personnel that had worked on the first refinery worked also on the second.

Training Iranians was an important part of Fluor's participation in the project. Iranian personnel were given formal courses in construction techniques such as welding, pipefitting, electrical and instrumentation work, as well as in design and engineering. The latter classes ran nearly three months before work began.

In 1975, Fluor formed a local engineering company, Fluor Iran, as a joint stock company, 51 percent owned by Iranians. It utilizes trained and experienced personnel from the temporary refinery engineering office. It is too early to say how the new joint company will fare. It seems certain that it will not be able to move ahead as fast as could a wholly controlled Fluor subsidiary. Not only does Fluor not have full, or even majority, control of the Iranian company, but the ability of the Iranian chemical engineering industry to absorb Fluor's sophisticated techniques will be a limiting factor.

It is interesting in this regard to contrast Fluor's performance on a much bigger refinery job for Mobil and that of the second NIOC job. Mobil engaged Fluor to design, engineer, procure, and construct one of the largest grass-roots[4] refineries ever built at Joliet, Illinois. The 160,000-barrel-per-day plant was fully on-stream in twenty-four months—and there was no construction overtime. The NIOC second phase, in contrast, took twenty-seven months to get most of the units on-stream, despite the fact that many of them were duplicates of the first phase and that the second phase was essentially a continuation of the first, and not a grass-roots job.

Fluor credits its success at Joliet to the "cooperation of Mobil, labor and Fluor," as well as to its management techniques. Clearly the absorptive ability of the environment where the job is performed affects its performance.

An aspect of the job environment that is of vital concern in considering the impact on the U.S. economy of technology transfers effected by the chemical engineering industry is the ability of the environment to meet the supply requirements for the job. In developing regions, few of the necessities are available locally, and U.S. chemical engineering companies will tend to favor the U.S. as a source as long as U.S. prices are competitive.

The $4 billion gas-collecting and processing job in Saudi Arabia will generate as much as $3 billion in equipment orders most of which will be filled in the U.S. The Saudi Arabian job may be exceptional in this regard, inasmuch as the Saudis have instructed Fluor to maximize the amount of equipment and the degree of engineering obtained in the U.S. U.S. prices—both for equipment and for engineering—are now very competitive, so the Saudi instruction is clearly self-serving. The arrangement is also in the best interests of the U.S. because it represents a substantial boost to the equipment industry and its suppliers. At least 50 percent of an equipment order will go into direct payments to the equipment suppliers' work forces or that of the subsuppliers. One can, therefore, translate the $3 billion Saudi equipment order into a substantial number of jobs in the U.S. If it is estimated that the average cost of a worker is $30,000 per year and that half of the $3 billion goes to pay labor costs, some 50,000 man-years will be needed to fill the Saudi order. The equipment will be supplied over some two to three years, so the order could be supporting some 20,000 workers during that period.

The remaining $1 billion of the Saudi contract will go into engineering and construction. With the instruction to maximize the amount of American involvement in these phases, it would not be far-fetched to say that 75 percent of the $1 billion would pay for U.S. services. U.S. engineers, construction workers, and supervisors probably cost a company like Fluor an average of $100,000 per year each.[5] The $750 million for engineering and construction, therefore, would translate into 7500 man-years of effort. As the total job will last some three to four years, one can say that more than 2000 U.S. engineering and construction personnel will be employed on the Saudi contract.

As was mentioned previously, the Saudi job is exceptional both in its size and in its instruction to maximize U.S. involvement. It does demonstrate, however, the potential for the U.S. chemical engineering industry to generate immediate returns to the U.S. economy for the relatively marginal cost of seeming to give away "vital" U.S. technology. As was also mentioned, the improvements to the technology made by putting it into practice, may soon make that which was transferred more obsolete than vital.

Notes

1. Individuals both within and outside the industry have recommended that companies explore the possibilities of jointly funded and operated research that would avoid unnecessary and costly competition and at the same time conserve scarce resources—particularly the highly skilled personnel needed for such work—to enable a new round of technological breakthroughs to be made.

2. Several contractors in the industry have made a plea for better government assistance in the course of conducting business, with aids such as increased access to government-supported finance, underwriting of high bidding costs and the price performance guarantees, and better working relations between the foreign representatives of the government and industry to improve commercial intelligence.

3. Fluor has recently been contracted to build 65 percent of yet another refinery for NIOC near Isfahan.

4. "Grass roots" is a term of art referring to construction work performed in an area where work of that type had not previously been done. Only grass roots existed before the work started.

5. The $100,000 would cover all costs, including Fluor's profit, and the high expense of maintaining U.S. personnel in remote and underdeveloped locations such as Saudi Arabia.

7

Policy Implications of Technology-sharing Trends

To the extent that the case materials are representative of conditions and trends, they pose new sets of problems and challenges for corporate management, the U.S. economy, and developing nations. The policy implications for each of these constituencies are analyzed in this concluding chapter. Table 7-1 provides an overview of corporate strategies versus purchaser objectives, as reflected in the case materials.

Challenges to Corporate Management

U.S. corporations now confront deep-seated external adjustment problems in their international marketing and production activities as a result of the profound changes that have occurred in the world economy over the past twenty-five years. Sector experience and corporate response, as depicted in the case material, are summarized in succeeding paragraphs.

1. *Aircraft Industry.* High R and D and production-tooling costs and nontariff barriers are forcing U.S. firms to share new aircraft design functions and to enter into coproduction agreements to reduce burdensome capital outlays for component production and to facilitate market entry. Defense considerations also have prompted U.S. firms to enter into technology-sharing and coproduction agreements.

2. *Automotive Industry.* Emergence of management services contracts in lieu of equity investment is a result of diminishing access to world markets, coupled with heightened risks of foreign capital investment and reinforced by the enhanced bargaining power of newly industrializing nations. The benefits accruing from cross-licensing and the proliferation of capital requirements (particularly where local borrowing is no longer feasible) also have led to technology sharing.

3. *Computer Industry.* Mounting R and D costs to develop new generations of products, the dearth of venture capital (particularly for smaller U.S. firms), and intensified foreign competition have forced some firms to withdraw from the market and to sell residual technology to foreign firms or release advanced technology to foreign enterprises in return for capital funding. Access to otherwise inaccessible markets (in Eastern Europe and other newly industrializing countries) have prompted coproduction and comarketing arrangements.

143

Table 7-1
Summary of Cases: Corporate Strategies versus Purchasing Group Objectives and Bargaining Leverage

Purchaser Group Objectives and Bargaining Leverage / Corporate Strategies	Industrially Advanced Nations	Socialist Countries[a]	Developing World Nations
	Government policy to develop internationally competitive industry in key sectors (e.g., computers and aircraft) / Public subsidy of R and D and production tooling / "Buy national" policies and nontariff barriers erected / High absorptive capability	State-negotiated transfer aimed at technological self-development / Coproduction and comarketing arrangements / Prepared to provide investment capital and pay for design-engineering costs / Need to export for foreign exchange earnings / Medium-level absorptive capability	National policies of rapid industrialization based on foreign technology / Seek technological self-sufficiency / Close internal market to exports / Willing to pay hard currency / Low absorptive capabilities
Explicit policy to shift from equity investment and management control to sale of technology and management services	Cummins-Komatsu (Japan) / Amdahl-Fujitsu (Japan)	GM-Polmot (Poland) / Control Data-ROM/SLA (Roumania) / RCA/Corning-UNITRA (Poland) / Sohio-PRC / Gamma-Socialist country	GTEI-SONELEC (Algeria)
Necessity to accept foreign affiliate or divest due to capital shortages, enormity of R and D outlay or production tooling requirement, nontariff barrier to market entry, or offset purchase requirements	GE-SNECMA (France) / General Dynamics-European Consortium / Honeywell-CII/HB (France)		Piper-EMBRAER (Brazil)
Measured release of core technology			Fluor-NIDC (Iran)
Sale of technology no longer considered central to company business or commercially advantageous	UOP-Auto firms (Japan) / Bendix-Bosch (W. Germany) / Motorola-Matsushita (Japan)		

Source: Developing World Industry and Technology, Inc.

[a]Refers to the nonmarket economies of Eastern Europe, the Soviet Union, and the People's Republic of China.

4. *Consumer Electronics Industry.* Increased foreign competition coupled with diminishing returns has forced multiproduct companies to phase out certain product lines and sell off technology to foreign firms. Traditional manufacturers have entered into remunerative management services contracts in newly industrializing nations (GTEI-SONELEC/Algeria) where market access would otherwise be blocked.

5. *Chemical Engineering Industry.* This industry is perhaps the most advanced (compared to the other four sectors) in terms of firms that specialize in the design, engineering, and construction of industrial plants for foreign clients. Such firms do not have the preoccupations of "product" companies, which are normally less willing to share technology with potential competitors. (The UOP case provides an interesting illustration of this distinction.)

Implicit in the industry trends described are several deep-seated challenges to traditional modes of global operation. There are also implications for adjustments in corporate planning in the foreseeable future.

Demands to Share Technology

Demands to share technology have come in large part from newly industrializing nations, but these pressures also derive from industrialized nations in advanced-technology fields (particularly aircraft and computers). In sharing technology, the problem is one of retaining core technology as a hedge against future competiton. This problem is particularly acute for the "product" company as distinct from the design engineering firm (see UOP case). It is less of a problem for the firm with a dominant technological lead and adequate earnings to fund R and D to maintain that technological edge.

In an ever-widening range of market circumstances, product companies in particular find themselves in the difficult dilemma of having to share technology or give up market positions. Corporate decisions to share technology are reinforced by changing corporate attitudes toward equity position and managerial control. On the one hand, there is the changing attitude that attractive returns can be earned from noncontrolled overseas commitments. On the other hand, the ever mounting political risks and economic uncertainties diminish the attractiveness of investments in overseas plant and equipment. The difficulties in maintaining proprietary rights and in remitting earnings from investments have reinforced U.S. corporate reluctance to make sizable financial commitments.

Technology sharing poses a new set of challenges for U.S. corporations. Unless the firm releases technology because it has decided to phase out of a particular product field, the measured release of core technology (the part that may represent a company's competitive edge) becomes critical. Consequently, it is necessary to define in operational terms what is unique, competitive, and proprietary—and, in fact, may be withheld.

Technology sharing generally implies profound changes in corporate management and organization. The transfer agents required to import and implant technology represent a quantum jump in operational requirements—moving from traditional marketing and manufacturing within the global corporate family to imparting these capabilities to enterprise partners through training and other enterprise-to-enterprise contacts.

Demands for World Marketing Assistance

A second interrelated set of demands from newly industrializing nations requires that U.S. corporations design, construct, and help put into operation industrial complexes that can compete in world markets. Such demands are not particularly problematic for firms such as Kellogg and Swindell-Dressler, which are exclusively in the design-engineering business. It does pose new dilemmas for product companies that are asked to design and engineer internationally competitive facilities and then assist the client firm to help market their exports (as in the General Motors-Polmot case). From the developing-nation viewpoint, this assures optimal economic efficiency in installed plants and equipment and, equally important, provides the means for earning much needed foreign exchange to help pay for the acquisition of highly sophisticated technology.

From the corporate viewpoint, the demand for assistance in world market entry may represent, as in technology sharing, a quantum jump in the logistical complexity of international operations. The overview of trends in the automotive industry describes the evolution in world involvements as the industry moved from marketing to world manufacturing and interchange systems. Specialized manufacture and interchange systems on a global scale, pioneered by IBM and others, have generally occurred within the centrally controlled framework of the multinational corporation. Indicative of the demands are the increasingly popular buy-back arrangements, whereby the seller of technology receives partial payment in product. These arrangements prevail with Eastern European countries in the computer equipment and consumer electronics fields and in manufacturing sharing with Japanese firms in the automotive (Cummins-Komatsu diesel engines) and computer equipment (Amdahl-Fujitsu) fields. Successful incorporation of buy-back arrangements into company operations requires a new order of integrated planning among the functional areas of marketing, production, design and engineering, research and development, and financial management. Production sharing also may pose some difficult problems of maintaining the quality and technical standards of trademark items (particularly when manufacturing facilities in newly industrializing nations are involved).

Buy-back arrangements, which at first were viewed as a necessary evil for market entry or retention of market share, are gradually coming to be viewed as a hedge against import restrictions and exchange controls and as a means for remitting earnings in areas where companies retain equity positions. In some cases, companies find offshore manufacturing of components and parts less costly than in the U.S. and in those cases have a double incentive to enter into buy-back arrangements. However, U.S. firms now face mounting opposition from labor unions because of the employment-displacing effects of these buy-back arrangements. Corporations thus find themselves in a cross fire of demands from newly industrializing nations and domestic constraints because of the possible adverse effects upon the U.S. economy.

Alternative Technology Sources

Since the late 1950s, competition from foreign technology sources (Japan and Western Europe) has intensified. As our overview of the chemical engineering industry indicated, although business has expanded tenfold over the past twenty years, the U.S. share of that business has dropped from nearly 100 percent to less than 50 percent in the hydrocarbon field (Chapter 6). For U.S. product companies, the increased competition means that they no longer wield the same bargaining leverage in entering foreign markets on their terms or in withholding their technology. There are very few product areas where nearly equivalent technologies are not available from an alternative source. If one U.S. firm refuses to share its technology with a potential purchaser, it generally faces the prospect that equivalent technology will be obtained from an alternative source (American or foreign), and the holdout company in any event will face a new competitor supplied by another source.

Several of our cases involved the sale of technology that was not or is no longer central to the firm's business (Motorola-Matsushita). In other cases the prospect of commercializing the product has proven difficult, and the U.S. firm has found offshore opportunities to commercialize the product (UOP-Japanese Auto and Bendix-Bosch). In cases involving a technically advanced and commercially strong foreign partner, rapid commercialization of the acquired technology may be executed.

The dearth of venture capital, coupled with the enormity of investment requirements in both R and D and production facilities, has also contributed to the spread of technology to foreign suppliers. Amdahl-Fujitsu in the computer equipment field and Cummins-Komatsu in the automotive product field (diesel engines) are two cases in point.

In most of the foregoing instances, companies have been releasing technology earlier in the product cycle than has been the practice traditionally. Until a decade ago, the traditional pattern was to release proprietary technology only after the product (and related production technologies) was "mature," that is to say home markets had been saturated and the outer reach to external markets through controlled investment had been exhausted. U.S. firms now face an intensification of competition in terms of the quantum and pace of technical change in an ever-widening range of industries and product groups.

Toward New, Commonwealth Arrangements

As a consequence of the described challenges that now face U.S. firms in the world economy—and demonstrated in the summary of sector experience—segments of U.S. industry have begun to move toward what may be termed "commonwealth" arrangements with sovereign foreign enterprises. The emerging commonwealth system traces back to the 1950s, when firms like IBM introduced specialized manufacture and interchange systems designed to overcome trade restrictions and foreign-exchange controls. Manufacturing responsibilities were apportioned to marketing areas so that export earnings could pay for required imports of finished products (electric typewriters and computer equipment in IBM's case) and components. Product and component specialization among participating countries permitted economies of scale in production and assured the necessary foreign exchange balances. The IBM system began in Europe and eventually spread worldwide with manufacturing and technical control rights for particular products assigned to individual countries.

IBM's specialized manufacture and interchange system has remained to this day under their full ownership and managerial control. IBM purchases materials and parts from suppliers with their own technology, but it shares its core technology with no one. A critical element in this global posture is the technological lead IBM has managed to maintain over would-be competitors. Other equally large companies in the automotive and consumer electronic fields, whose technological leads have been eroded by competing enterprises at home and abroad, have been edged out of certain product lines or now feel compelled to share technology in order to maintain segments of world markets.

Technology sharing thus represents another quantum step in corporate evolution. It portends further deep-seated changes in the organization and logistics of world production, marketing, and, ultimately, the design and engineering of industrial systems. Certain products will have to be designed and production-engineered to accommodate not only the multiplicity of markets they may have to serve but also their specialized manufacture and interchange among sovereign enterprises. These adjustments have far-reaching implications for global corporate planning and logistics and also carry with them labor-market adjustment problems for the U.S. economy and opportunities for newly industrializing nations.

Impact on the U.S. Economy

The evidence presented in the case material is an early warning of troublesome trends in the international economy: an enlarged mobility of technology in international trade, the importance of technological diffusion in changing patterns of comparative advantage, and intensified competition among suppliers of technology. In the U.S. economy, the mounting costs of R and D and, in some cases, the difficulties in obtaining venture capital, have contributed further to the release of competitive technology to foreign enterprise, the build-up of competitive production capabilities abroad, and the narrowing of international gaps in high-technology industries.

Other factors contributing to technology-sharing trends have been the profound changes in corporate philosophy on the management of technology assets. An increasing number of firms have found technology sharing highly lucrative and reinforcing of corporate strategies aimed at maintaining global market positions and earnings. This tendency is evident in the shifting emphasis among many companies away from investment in production facilities and toward increased expenditures on R and D and marketing in order to maintain global positions rather than competitiveness in the U.S. economy. There is mounting evidence that U.S. firms are faltering in their will and determination to continue to design and engineer for the high-wage U.S. economy.[1] Once productivity begins to lag behind wage increases, U.S. industry loses its competitiveness in the world economy. If those trends reach significant proportions, they will unleash a series of deep-seated adjustment problems for the U.S. economy.

Job Erosion

In the technology-sharing arrangements described in the case materials, the U.S. economy does benefit from the sale of design-engineering services involving U.S. personnel and purchases of U.S. equipment associated with overseas plant construction and from the export of materials and components during plant run-in periods. There may even be follow-on sales of goods and services in related product lines once the core technology has been established abroad. It is when the internationally competitive facility is on-stream and begins to export competitive products or componentry, reinforced and facilitated, at times, with buy-back agreements, that U.S. production employment is adversely affected.

In a dynamic economy, job displacement is inevitable, and the problem is one of offsetting the decline of one industry with job generation and labor market mobility to maintain overall levels of high employment. Technological displacements have proven particularly troublesome under adverse domestic economic conditions (low growth rate, high unemployment, inflation, declining productivity, and balance-of-payment difficulties), in the absence of substantial improvement in labor-market adjustment mechanisms (relocating and retraining displaced production workers), and in an economy where technologically dynamic industries (or services) are not expanding at a rate sufficient to absorb labor displacement from declining industrial sectors. The shift from the low-skill range of production jobs to higher-skill technical support poses additional manpower adjustment problems.

For some time, conventional wisdom has argued that U.S. industry has a self-interest in restricting the release of proprietary technology and that that maxim provides a sufficient safeguard of U.S. economic interests. Whether the latter part of that statement can continue to be valid depends on whether American firms maintain their commercial leads in high-technology fields, whether these firms continue to invest in R and D to maintain their technological leads and continue to be the major sources of design and manufacturing technology, and whether the industrial dynamics of other economies (particularly in Japan) do not outpace our own efforts to maintain competitiveness in the world economy.

The case studies and our assessment of the competitive position of technologically based U.S. industry in the world economy indicate that these necessary conditions do not at present and will not in the near future prevail. This situation is a particularly disturbing portent for the U.S. economy. In addition, there seem to be new dimensions in the nature and extent of the outward movement of technology that may be taxing the limits of our economic adjustment capacities.[2] The release of the most advanced, sophisticated, and hence competitive technologies, particularly to commercially astute and aggressive foreign enterprise can be especially destructive to U.S. production jobs. The Amdahl-Fujitsu and Cummins-Komatsu agreements are once again cases in point. In the former instance, the Japanese partner is already in the process of licensing acquired computer technology to Western European partners and is thereby outflanking U.S. access to that market with either products or technology.

It is also worth noting that U.S. production jobs are now threatened from three distinct sources, beginning with the technology-sharing arrangements that have been described in the case materials. Second, the off-shore manufacturing operations of U.S. corporations seeking to take advantage of low-wage, high-productivity foreign labor, have a negative effect on U.S. production jobs. The third, newly emerging threat comes from investments by foreign corporations in U.S. assembly facilities in the automotive, electronics, aircraft, and steel finishing industries.[3] At first blush, these industries generate jobs, but to the degree that they are used as assembly operations, a large percentage of the componentry and semi-finished materials are imported from abroad. Thus an 85 percent Japanese-content Honda may displace a 90 percent U.S.-content American compact car.

Design-engineering for the U.S. Economy

There is some evidence that U.S. firms are encountering increasing difficulties in adjusting to technical change and are considering marketing their technology as an alternative to aggressively engineering for competitive production in the U.S. economy. In the Cummins-Komatsu case, the corporate decision was to allocate major manufacturing responsibility for its most advanced engine line to a strong, successful Japanese firm (formerly a licensee) because of the faltering proficiency of production in the U.S. and the cyclical difficulty of raising expansion capital in the U.S. The dearth of venture capital was a determining factor in the substantial release of front-end technology in the Amdahl-Fujitsu case.

In October 1977, the Zenith Corporation, one of the last holdouts among large American corporations in the consumer electronics equipment field retaining the bulk of their manufacturing in U.S. plants, finally threw in the towel and announced it would move offshore for most of its television assembly operations and would discontinue manufacture of stereo equipment.[4] The new generation of video-tape recorders—projected to be a $1 billion industry within three years—is destined from the outset to be a Japanese industry. Zenith is licensing technology from SONY and RCA has teamed up to market Matsushita's products. The fact is that Japanese firms have outdistanced American in economical design-engineering of componentry in this and other product fields.[5]

The proliferation of internationally competitive, industrial technology to Japan, Western Europe, and socialist economies may be weakening the bargaining position of U.S. firms as suppliers of technology to newly industrializing countries and in trade negotiations with these countries. Japan has become a technological intermediary for many countries of the developing world, having absorbed and adapted segments of U.S. technology. The socialist countries may also assume a similar role if the current transfers are of sufficient volume and are efficiently implanted. These trends in the world economy add to the continuing necessity for U.S. industry to maintain technological leads through investments in R and D and the commercializing of technical innovation *within the U.S. production system* (as distinct from the tendency to move abroad) as it becomes difficult to maintain production competitiveness in the U.S. economy.

Public Policy Versus Corporate Interests

Discussions of public policy issues as they relate to technology and corporate planning are complex and difficult for several reasons. To begin with, there is the heterogeneity of issues that beset this area. International technology transfers have given rise to a broad array of intricate public policy issues ranging from industrial employment erosion effects to impact on national security.[6] The former links the sale of proprietary technology to foreign enterprises to the possibility of a decline of U.S. competitiveness in world markets as rival economies begin to compete for markets heretofore supplied from U.S. production sources. The set of issues associated with national security considerations is compounded by the potential threat of enhancing the technical and economic proficiency of a potential military rival through the release and implantation of U.S. technology.

Second, there are difficult semantical problems in this area, even problems of basic literacy among involved parties. Terms such as "technology" and "transfer" are often the source of confusion and misunderstanding between business and government communities at home and abroad. Government officials treating technology in simplistic terms can play havoc with corporate managers responsible for managing intricate technology systems.[7]

There are other critical distinctions to be drawn between the transfer of a turnkey facility and associated operational capabilities, on the one hand, and the additional knowledge and training, on the other, that contribute to implanted capabilities for duplicating similar industrial facilities without additional outside assistance and even beyond that to design and engineer future generations of industrial systems.

A third factor that compounds government-industry relations in technology policy areas stems from the wide diversity in the backgrounds and experience of corporate representatives compared to those concerned with public policy. The latter often include academicians with theoretical orientations rather than practitioners, who have very different perceptions of problems and issues. In addition, the financial, legal, economic, commercial, and technical dimensions of the problems very often bring together people of very different professional backgrounds.

The foregoing difficulties give rise to a very special set of conflict resolution problems. The "actors" on this "stage" often have highly diversified perceptions of issues, and they also may represent conflicting sets of objectives and interests. The economic interest of the U.S., in terms of employment and labor market adjustment, may conflict with the commercial interests of a particular firm selling its technology to a foreign entity. Similarly, a foreign government may consider it necessary or desirable to limit the equity position and managerial control of a U.S. firm in order to optimize national enterprise development.

There is also the problem of consensus among interest groups. Within the corporate world, there are often considerable differences between hardliners who have a very restrictive set of policies regarding technology sharing with foreign enterprises and firms that have a more liberal policy of technology sharing with associated firms overseas. The former typically are technology leaders and have a strong market position, including financial resources to capitalize their technology assets—in contrast to firms that do not have a predominant market position.

A similar diversity of viewpoint may be found within the U.S. government. Defense-related agencies are generally adverse to the release of technology to Communist countries and other potential adversaries. The Departments of Commerce and State generally favor free trade in technology for commercial or diplomatic reasons. Treasury, because of its concern with balance of payments and tax incidence often takes an interventionist position where the foreign activities of U.S. firms are concerned. The Department of Labor is increasingly concerned over U.S. job losses resulting from the release of U.S. technology.

Opportunities for Newly Industrializing Nations

Development authorities and enterprise groups in newly industrializing countries may gain several important insights from the case materials contained in Chapters 2-6. The inclinations of U.S. corporations to move in the direction of commonwealth arrangements provide an array of opportunities to develop technologically and to realize a number of their articulated goals within the framework of a new international economic order. An important caveat to these new opportunities is in order, however. These very proclivities have contributed to the outward movement of production technology that threatens U.S. employment levels, and this tendency may provoke a protectionist response within the U.S. Such a reaction, in turn, would inhibit the expansion of industrial exports from developing countries. Successful movement toward a new international economic order is tied to a harmonization of trade, development, and commercial interests of the involved governments and enterprises.

Toward a New International Economic Order

The technology factor in national efforts toward rapid industrialization is crucial. For newly industrializing nations, it is not just a matter of acquiring turnkey plants that they can operate. Development authorities are increasingly aware that the key to long-term growth and development lies in acquiring the ability to design and engineer industrial systems. This requires an indigenous build-up of an interrelated set of capabilities, including product design and engineering process and equipment design and construction. It is for this reason that developing world nations have been rejecting ownership and managerial control of national enterprise by foreign corporations. Foreign enterprises bring with them industrial technology and obviate the need for developing indigenous design and engineering capabilities or supporting capital goods industries. Once established in an economy, they preempt the development of indigenous industry because of their comparative advantage in management and technology.[8]

To develop indigenous design and engineering capabilities, developing world nations need to train engineering and technical personnel, upgrade the technical absorptive capabilities of industrial enterprises, and develop supporting technological infrastructure in design and engineering groups and capital goods industries. The technology-sharing agreements described in the case studies indicate how developing world nations may take advantage of the new relationships into which multinational corporations and other foreign enterprises now seem willing to enter. (See, in particular, GTEI-Sonelec/Algeria, Piper-Embraer/Brazil, Control Data-ROM-SRL/Rumania, and General Motors-Polmot/Poland.)

The bargaining position of developing world enterprise depends on the economic environment in the country (including size of the internal market and access to third-country markets), the financial position of the enterprise and the economy to fund technology acquisition without foreign equity (oil-rich economies such as Iran or Algeria, for example), the role of government (in screening and controlling foreign investment and licensing or in financing both design and engineering and capital investment components), and the technical absorptive capabilities at the enterprise and sector levels (including astuteness in developing alternative technology sources and in "unbundling" technology into process, equipment, construction, and training components).

Technology Acquisition Strategies

The techology-sharing agreements between U.S. firms and various socialist countries, involving production-sharing and comarketing arrangements, respond to a number of the new development goals aimed at selective and progressive technological self-reliance. They provide for the rapid implacement of internationally competitive facilities (to assure economic efficiency and to earn much needed foreign exchange) and for the training of managers and engineers (to take full operational control as rapidly as possible). In Brazil, this type of agreement is assisting the infant aircraft industry at a critical phase in its efforts to become internationally competitive (see Piper-Embraer/Brazil). Foreign firms can provide the overall program and logistical support to help develop the indispensable component and materials supplier industries in automotive, aircraft, and electronic equipment manufacture.

The case material also illustrates various tactics that developing world governments may consider for the support of national technological development. The protection of local industry from foreign imports and from the competitive presence of multinational corporations during the infant stage of development is indispensable, as the Japanese experience has demonstrated. Government intervention in high-technology industries has also proven necessary in industrially advanced countries in the jet aircraft (General Electric-SNECMA/France and General Dynamics-European Consortium) and computer equipment fields (Amdahl-Fujitsu/Japan and Honeywell-CII-HB/France).

If a newly industrializing country is to develop indigenous design and engineering capabilities, national laws limiting the activities of foreign design-and-engineering firms may be necessary (U.S. firms are prohibited *de facto* from practicing in Japan). Collaborative efforts with U.S. design-and-engineering firms afford a wide range of training and development opportunities (Sohio-People's Republic of China and Fluor-Iran and Saudi Arabia). In some cases, the developing-world affiliate can become part of the foreign firm's international bidding network. In certain areas, developing-world enterprises may be able to take advantage of corporate strategies to phase out peripheral technologies (Motorola-Matsushita/Japan) or where the firm is capital short and is seeking a competent technical partner for its global markets (Cummins-Komatsu/Japan).

Technological partnerships between U.S. corporations and industrial enterprises in developing nations can be mutually beneficial. Until such time as the technology and managerial gaps are considerably narrower, U.S. enterprises can continue to export U.S.-manufactured goods, engineering services, and capital equipment. In this respect, U.S. trade in technology with developing world nations should stem from a substantially different policy from the one for Japan and other industrialized economies, where the technology gap has been narrowed or no longer exists. In a changing world economy in which developing countries take an expanding share in the international division of labor for manufactured goods, the U.S. will continue to shift to the capital-intensive range of manufacturing and to relinquish the more labor-intensive segments of production to lower-wage countries. In one respect, the earnings generated by technology sales could result in new and more competitive industry, and the newly industrialized nations that benefit from the purchase of our technology could expand demand for U.S. exports. These shifts, nevertheless, place an added burden on U.S. labor-market adjustment mechanisms.

Tables 7-2 through 7-6 (pages 158-163) present the major features of the case studies grouped by industry. In addition to describing the nature of the agreement and the principal considerations of the technology supplier and purchaser, the tables outline, in the final column, the policy implications derived from the case studies for the three constituency sets analyzed in this chapter: corporate management, the U.S. economy, and the newly industrializing nations.

Notes

1. See chap. 1, n. 2.

2. See, for example, Harry G. Johnson, "Technological Change and Comparative Advantage: An Advanced Country's Viewpoint," *Journal of World Trade Law* (January-February 1975):1-14.

3. For an excellent discussion of this phenomenon, see Yoshi Tsurumi, *The Japanese Are Coming: A Multinational Interaction of Firms and Politics* (Cambridge, Mass.: Ballinger Publishing Co., 1976), chap. 4.

4. "Zenith to Lay Off 25% of Workers Within a Year." *Wall Street Journal,* 28 September 1977. "Zenith Losses Force Shift Overseas." *Washington Post,* 29 September 1977. The *Post* article notes that Zenith has been one of the most determined and successful U.S. firms in its effort to retain its U.S. workers.

5. Peter J. Schuster. "A $1300 Christmas Toy Made in Japan." *Fortune,* September 1977. See also Tsurumi, *The Japanese Are Coming,* chap. 7, for an analysis of the Japanese design-engineering capabilities.

6. The former was dealt with in "International Transfers of Industrial Technology by U.S. Firms and Their Implications for the U.S. Economy," prepared by Developing World Industry and Technology for the Office of Foreign Economic Research, Bureau of International Labor Affairs, U.S. Department of Labor, December 1976; the latter was the concern of *An Analysis of Export Control of U.S. Technology—A DOD Perspective,* a report of the Defense Science Board Task Force on export of U.S. technology, February 1976.

7. Technology in an operational sense is more than a collection of paper recipes. It represents an intricate set of detailed knowledge, skills and

specifications on end-product designs, materials, factory methods, materials processing techniques, plant layouts and designs, equipment design and utilization models, and management and control procedures. These intricate "knowledge" sets are embodied in people, equipment, blueprint manuals, and other written materials. Industrial technology is usually imparted or transferred through sustained enterprise-to-enterprise relationships during which time the detailed knowledge sets are conveyed by the written and spoken word, procedures emulated through trial and error, skills imparted through training efforts, and product designs or production techniques reworked to fit an anomalous set of conditions.

8. It is for this reason that the Japanese government passed laws in the early 1950s screening all foreign investment and controlling foreign licensing arrangements to assure that technology was acquired on terms and conditions favorable to the rapid growth and development of Japanese enterprise.

Table 7-2
Case Summaries and Overview of Policy Implications: Aircraft Industry

Case Studies	Technology Transferred	Corporate Considerations	Purchaser Considerations	Policy Implications
General Electric-SNECMA (France)/joint venture	Design and production of high technology, advanced-design jet engine for civilian aircraft	Fear of being precluded, even on a partial basis, from lucrative European market	Necessity to accept strong foreign partner to obtain internationally competitive technology Strong backing of French Government to acquire design and engineering	*Corporate Planning* Foreign firm(s) assisted to point of becoming major innovator and competitive threat to U.S. industry (GE-SNECMA, General Dynamics-European Consortium)
General Dynamics-European Consortium/Coproduction agreement	Production technology for advanced fighter aircraft	Sale of technology helps amortize huge R and D costs Need to maintain technological lead to remain internationally competitive	Insistence by purchaser government on coproduction and offset purchase arrangements	*U.S. Economy* Progressive erosion of U.S. technological lead as result of release of advance technology systems and imparting design-engineering capabilities (GE-SNECMA, General Dynamics-European consortium)
Piper Aircraft-EMBRAER (Brazil)/licensing agreement	Production technology for light civil aircraft	Need to maintain technological lead and production competitiveness against future Brazilian export capabilities	Piper preempted from export by Brazilian import restrictions Brazilian government interest in developing national design and production capabilities	Prospect foreign associate may become low-cost producer of product (Piper-EMBRAER) *Newly Industrializing Nation (NIN)* Opportunity for NIN enterprise to enter into world market on competitive basis—including upgrading of industrial design and engineering capabilities (Piper-EMBRAER)

Source: Developing World Industry and Technology, Inc.

Table 7-3
Case Summaries and Overview of Policy Implications: Automotive Industry

Case Studies	Technology Transferred	Corporate Considerations	Purchaser Considerations	Policy Implications
GM-Polmot (Poland) coproduction agreement	Design and production for new line of commercial trucks to be marketed internationally	Opportunity to earn corporate return on Eastern European and segment of Western European markets that would otherwise be, respectively, inaccessible or require prohibitively high capital investment	Polish government wanted training of technicians and managers in production design operations as well as in engineering. Polmot wanted rapid placement of internationally competitive facility to earn foreign exchange	*Corporate Planning* Future emphasis in U.S. enterprise on marketing and product research and internationalizing production function (Cummins-Komatsu)
Cummins Engine-Komatsu (Japan)/licensing agreement	Manufacturing technology for advance design diesel engine	Severe capital constraints prompted company to assign major manufacturing responsibilities to former Japanese licensee	Komatsu anxious to take on major manufacturing role for world market	*U.S. Economy* U.S. firm contributes to export competitiveness of foreign manufacturer and potential job erosion in U.S. production base (General Motors-Polmot, Gamma Auto-Socialist Country
Gamma Auto-Socialist country/licensing agreement	Manufacturing technology for latest generation automotive part	Corporate plan is to develop next generation of manufacturing technology that will cut costs by 40% before purchaser is ready to enter and compete in international market	Purchaser government anxious to obtain most modern and up-to-date facility to manufacture auto components to internationally competitive standards	Cross licensing arrangements reinforce competitive position of foreign firms vis-a-vis other U.S. firms (Bendix-Bosch)
Bendix-Bosch (Germany)/cross-licensing agreement	Know-how to manufacture electronic fuel injection system	Necessary to license front-end technology to strong technical partner with dominant market position in order to earn return, to avoid patent infringement claims, and to benefit from technology exchange	German firm able to maintain technological parity with U.S. firm through cross-licensing arrangement. European auto market demand ahead of United States for this technical innovation	*Newly Industrializing Nations (NIN)* Opportunities for NIN enterprise to enter into coproduction, comarketing arrangements (GM-Polmot)

Source: Developing World Industry and Technology, Inc.

Table 7-4
Case Summaries and Overview of Policy Implications: Computer Industry

Case Studies	Technology Transferred	Corporate Considerations	Purchaser Considerations	Policy Implications
Amdahl-Fujitsu (Japan)/ licensing agreement	Manufacturing technology for advanced generation computer mainframes used in conjunction with IBM peripherals and software	Inability to obtain venture capital in U.S. market led to association with Japanese firm	Japanese firm anxious to obtain internationally competitive technology to compete in Japanese and other world markets Strong backing of Japanese government	*Corporate Planning* Acquisition of U.S. product design poses competitive threat in third-country markets (Amdahl-Fujitsu) Foreign enterprise may use acquired technology plateau to design and engineer next generation technology (Amdahl-Fujitsu)
Honeywell-CII-HB (France)/joint venture	Manufacturing technology and computer design inherited from acquisitions of GE computer division	Sole means of obtaining access to French market Guaranteed sales to French government and heavy R and D support for technical development	Financial backing of French government to develop French computer industry to challenge dominant IBM position Anxious to obtain advanced generation of U.S. technology	*U.S. Economy* Enhanced competitive position of foreign firms in worldwide markets may lead to displacement of U.S. production jobs (Amdahl-Fujitsu, Honeywell-CII-HB)
Control Data-ROM-SRL (Romania)/ joint venture	Manufacturing technology for computer peripherals	Access to Eastern European market Low-cost procurement source for components exported to Western Europe	Romanian government interested in coproduction agreements that upgrade technological capabilities and earn foreign exchange for economy	*Newly Industrializing Nations (NIN)* Opportunities for NIN enterprise to enter into coproduction, co-marketing arrangements (as in Control Data-ROM-SRL)

Source: Developing World Industry and Technology, Inc.

Table 7-5
Case Summaries and Overview of Policy Implications: Consumer Electronics Industry

Case Studies	Technology Transferred	Corporate Considerations	Purchaser Considerations	Policy Implications
GTEI-SONELEC (Algeria)/turnkey contract	Plant construction, manufacturing technology (fully integrated from raw material to radio and TV) and training of full-range of managers and technicians	Opportunity to earn substantial return on technology assets without capital resource commitment	Oil revenues permitted government to finance industrial facility. Want self-sufficient operational and eventually design engineering to compete in world market	*Corporate Planning* Global prospect that astute industrial enterprise will use acquired technology as plateau from which to design and engineer future product systems (Motorola-Matsushita)
Motorola-Matsushita (Japan)/divestiture	TV manufacturing plant and patents sold to Matsushita	U.S. firm decided to phase out its involvement in consumer electronics field in face of mounting R and D costs and intensified competition from foreign (Japanese) imports. Divestiture came at time when consumer demand had fallen off in U.S. market	Matsushita gained some manufacturing know-how (less in product design). Company able to expand product line worldwide and rapidly penetrate U.S. market	*U.S. Economy* Release of technology to foreign partner(s) with high technical absorptive capabilities leads to rapid development of international competition (RCA/Corning, UNITRA). Erosion of U.S. production jobs occurs when foreign assembler in U.S. begins to displace U.S. component content (future possibility in Motorola-Matsushita type of situation)
RCA/Corning-UNITRA (Poland)/licensing agreement	Product design, plant engineering, manufacturing know-how, and technical training for cathode-ray tubes.	Financially remunerative opportunity to earn return on technology asset in otherwise inaccessible market	Polish enterprise interested in internationally competitive production capability to supply domestic and Eastern European market	*Newly Industrializing Nations (NIN)* Management service contracts provide opportunity for NIN enterprises to develop an internationally competitive industry, including training of vital technical and managerial personnel (GTEI-SONELEC, RCA/Corning, UNITRA)

Source: Developing World Industry and Technology, Inc.

Table 7-6
Case Summaries and Overview of Policy Implications: Chemical Engineering Industry

Case Studies	Technology Transferred	Corporate Considerations	Purchaser Considerations	Policy Implications
UOP-Foreign auto firms with U.S. export markets/licensing agreement	Catalysts and catalyst containers for catalytic converters	Process company focuses on the development and sale of successive generations of innovative technology; product firm, the Automotive Division, retains control over technology and manufacture and returns profit factor to company for each unit sold	Foreign auto makers with U.S. export markets compelled by law to install emission control devices to maintain market share	*Corporate Planning* Important distinctions between "product" and "design-engineering" companies in the logistics of developing successive generations of technologies and marketing them to other enterprises (domestic and foreign)
Sohio-People's Republic of China/licensing agreement	Process technology for synthetic fiber feedstock (acrylonitrile) Through Japanese engineering intermediary includes plant engineering and training of engineering personnel	Acting here as a process engineering firm (as distinct from product firm with concern over world future market shares), prime consideration is maximizing revenues from technical innovation	PRC anxious to acquire up-to-date operative technology with export potential and at the same time lay base for adaptive engineering of related families of chemical facilities	*U.S. Economy* Chemical engineering industry stands as a model to the measured release of industrial technology with refurnishment of stock of technology through R and D funded by corporate returns (Sohio-PRC, Fluor-Iran/Saudi Arabia)

163

Table 7-6 (cont.)

Case Studies	Technology Transferred	Corporate Considerations	Purchaser Considerations	Policy Implications
Fluor-Iran and Saudi Arabia/turnkey contracts	Engineering and construction of chemical and petroleum processing facilities, with a strong emphasis on efficient and rapid installation (Saudi Arabia—a natural gas collection plant; Iran—a series of refineries	Process and plant engineering firm's basic interest in maximizing returns from sale of design and engineering services anywhere in the world (over half of $10 billion in contracts now overseas) In order to maintain commercial and technological lead in world, firm must improve its cost competitiveness by constantly upgrading its efficiency in project management	Saudi Arabia anxious to expand its domestic processing of oil and natural gas resources Strong drive in Iran to develop local expertise in project management	Potential threat to U.S. production jobs comes from internationally competitive facilities producing commodities for world markets (Sohio-PRC, Fluor-Iran/Saudi Arabia) Safety and energy conservation regulations can have important impact on relative competitiveness of U.S. industry (UOP-Foreign Auto) *Newly Industrializing Nations (NIN)* NIN enterprises can negotiate contracts that move beyond turnkey operations into self-sustaining design engineering capabilities (Sohio-PRC)

Source: Developing World Industry and Technology, Inc.

Index

165

About the Author

Jack Baranson is an economic consultant to industry and government with a worldwide reputation in the field of international transfers of technology. He is president of Developing World Industry and Technology, Inc., a policy research and consulting group. Two major policy studies recently completed are *The Transfer of Industrial Technology by U.S. Corporations and Their Implications for the U.S. Economy* (for the U.S. Department of Labor) and *North-South Technology Transfer Issues: Mexico, Brazil, Venezuela and Colombia* (for the U.S. Department of State).

Dr. Baranson was a visiting lecturer at the Harvard Business School (1972) and served for six years as a staff economist of the World Bank (1965-1971). He has previously served as a research associate with the Brookings Institute, the Committee for Economic Development and the International Development Research Center at Indiana University. Dr. Baranson holds a doctoral degree in economics from Indiana University (1966) and a master's degree from the Johns Hopkins School of Advanced International Studies (1956). Dr. Baranson is the author of *Manufacturing Problems in India* (Syracuse University Press, 1967); *Industrial Technologies for Developing Economies* (Frederick A. Praeger, 1969); and *Automotive Industries in Developing Countries* (World Bank—Johns Hopkins Press, 1969).